# The Ghost Lake

The True Story of Louis Agassiz's
Daring Scientific Explorations

John Hudson Tiner

## The Ghost Lake

All rights reserved. Except as permitted under the U.S. Copyright Act of 1976, no part of this publication may be used or reproduced, distributed, or transmitted in any form or by any means, or stored in a database or retrieval system, without permission.

## This Book is Dedicated to John Watson Tiner

## The Ghost Lake - The True Story of Louis Agassiz.

Louis Agassiz convinced a reluctant scientific world that glaciers moved. A great Ice Age had once wrapped much of the world in a shroud of ice. He also discovered that, when melted, the ice left behind the Great Lakes.

Even when he was young, Louis Agassiz was fascinated by the world of nature. All creatures interested him. As he grew older, Louis learned all he could about animals and plants. He found fossils especially fascinating. He soon became well-known for his great knowledge of science.

He lived an exciting life. He began difficult scientific explorations. He climbed mountains, organized expeditions into unexplored areas, and had himself lowered into the heart of a glacier through a great crack in the ice.

His great knowledge of nature earned him the title of Nature's Librarian.

Table of Contents

Chapter 1: The Book of Nature
Chapter 2: Follow a Dream
Chapter 3: The Great Baron
Chapter 4: Animals in the Ice
Chapter 5: Doctor Louis
Chapter 6: Rivers of Ice
Chapter 7: Hotel on a Glacier
Chapter 8: The Heart of the Glacier
Chapter 9: Victory Over the Jungfrau
Chapter 10: A New Land
Chapter 11: Ghost Lake
Chapter 12: Nature's Librarian
Chapter 13: Desolate Theory
Chapter 14: Louis Agassiz Today

Chapter 1: The Book of Nature

    One early morning in the fall of 1820, Louis Agassiz and his young brother Augustus waded into the cold waters of Switzerland's Lake Morat. The surface of the water was mirror smooth.
    The two boys stood still as statues. Louis peered into the water. Nothing moved except for his alert eyes until he discovered a fish darting through the water. As the fish turned, light flashed over its bright colors. Augustus was tired of looking for fish. "Let's go, Louis. It's time for breakfast."
    Louis didn't move, but whispered, "Patience and quickness, Augustus."
    Augustus sighed. The water was cold and his legs were numb.
    Suddenly, with lightning quick hands Louis reached into the water and grabbed a brightly colored fish. It fought to escape him, flipping its tail back and forth. But Louis held on.
    "I've got him," he yelled. "Look at his colors! We don't have any like this."
    Augustus was unimpressed. "Let's eat," he said.
    "We've got to feed the animals first," Louis said.
    The boys waded ashore and raced barefoot toward home. Because their father was a minister they lived in a parsonage, but it looked no different from other Swiss homes with its small windows and low roof.
    In its front yard stood a great stone basin cut out of a granite boulder from the Alps. A wooden square pipe lashed together by leather bands carried water from a hillside spring to the stone basin.
    Louis dropped the fish into the basin and waited to see if it would swim. It darted back and forth unharmed by its brief life out of water.

Louis reached for feed in a bag by the side of the pool and tossed some to the other fish in the basin. The water rippled with movement as they surfaced for food. The Agassiz yard swarmed with animals. Farm animals including sheep, chickens, a rooster, a cow wearing a bell, and a sleepy-eyed horse roamed about, but wild animals lived there too. Some ran free, but others were confined. With their own hands Louis and his brother had constructed cages for mice, rabbits, birds, guinea pigs, and even snakes. In the backyard under an apricot tree hung a beehive.

As Louis visited each cage, his animals greeted him with squeals and squeaks. He fed them all and talked to them, calling each by name.

Louis's sister Cecelia came outside as he was caring for his animals. "I fed the animals for you while you were gone to the academy," she said. "Will you be leaving again for school after harvest?"

Louis paused. Reluctantly he closed the last cage.

"Father wants me to choose a Practical Profession such as bookkeeping or medicine."

Suddenly Louis spotted movement in the grass alongside the stone wall of the house. He ran toward it and reached down for a snake. He held it inches from his face and examined it in detail.

Cecelia shuddered. "I don't understand your fascination with snakes. They smell, and they're slimy."

Louis smiled, teasing her. "Snakes aren't slimy. Feel it. Its skin is cool and dry." He pushed the snake closer to his sister.

Cecelia moved away, suddenly finding flowers nearby that were more interesting.

Louis realized his sister wouldn't be teased any further and released the snake. "I'll wash for breakfast," he said.

Every meal at the Agassiz home was a family affair. Louis, Augustus, sisters Cecelia and Olympia, and Mother and Father Agassiz sat by the table. In the kitchen, their housekeeper rattled pots and pans as she prepared breakfast.

Father Agassiz bowed his head. "Let us ask a blessing for this meal."

Everyone became silent.

Father Agassiz prayed, "Heavenly Father, You are the maker of all creation and provider of all that we have. We ask for Your blessings upon us in this harvest season. Bless this family that we may do Your work."

"Amen," all said the others together.

After breakfast, Louis went to his room for a sketch pad. He wanted to draw in color a picture of the fish he had caught that morning.

When he stepped outside, a man in a fishing boat on the lake distracted him. "Do you want to come along?" the man yelled.

Louis's face lit up. "Oh yes, at least until the harvest horn blows."

In exchange for the ride on the lake, Louis knew the fisherman expected him to help with his nets. Many times they set out the nets and drew them in, but each time the nets came back empty.

Finally, the fisherman said, "We've got something. The nets are heavier."

Louis pulled hard, but instead of fish, one net held two old vases of blue pottery. Louis freed the vases from the net. "What are these?" he asked.

"They're a nuisance, just something to foul the nets," the fisherman said. "Our nets drag them up from the bottom of the lake."

"You've found others?" Louis asked.

"Hundreds of them," the fisherman said, "perhaps thousands."

Louis examined one of the clay pots. It looked very old, its soft blue color quite unlike anything he had ever seen before. On its sides were imprints in delicate designs.

"Who made these?" Louis asked.

The fisherman shrugged. "Nobody knows," he said, "or cares."

The old man picked up the other vase and hit it against the side of the boat. As it shattered blue pieces of pottery fell all about them. "That's one vase that won't foul our nets again!" He pushed its broken shards overboard and reached for Louis's vase.

Louis held on tight. "Not this one. I want it for my collection."

The fisherman said impatiently, "Is collecting things all you ever think about?"

"Why, no," Louis said. "There are so many things I want to do. I'd like to sail on a lake so wide you couldn't see land in any direction. I want to climb the Jungfrau to see for a hundred miles - " He pointed to the snow-capped peak, one of Switzerland's most famous mountains.

The fisherman smiled. "Climb the Jungfrau? That's extremely dangerous. The only men foolish enough to climb that high are chamois hunters. Most of them fail to come back, and even those who do return are injured in falls and the intense cold."

The chamois was a mountain animal that was as sure-footed as a mountain goat. Hunters prized its soft skin that could be sold for large sums of money. Chamois hunters made good money, but not without tremendous risks.

Louis asked if the old man had ever talked to chamois hunters.

"Many times," the fisherman said proudly. "The hunters tell stories about those peaks. There's an enormous glacier there with solid ice thousands of feet thick. It has cracks in it hundreds of feet deep into which a man could fall and never be seen again. Boulders are scattered in all directions as if thrown by a giant hand. Fog and snow constantly shroud the peaks. A careless step up there can lead to icy death."

Louis was absorbed in his own dream of exploring the mountain and muttered softly, "I'll go there. I know I will."

"The son of a simple preacher talks mighty big," the fisherman said roughly. "Shouldn't you enter your father's profession?"

"I might become a preacher," Louis said. "But there is more than one way to study God's word. He wrote the book of nature too. I want to be a scientist."

The fisherman looked disgusted. "Scientists stuff their brains with useless book learning."

Louis argued, "You don't learn science from books. You have to explore nature to study it firsthand." The long, high call of an Alpine horn sounded across the lake.

"I have to go," Louis said. "That's the call to harvest. We have to pick grapes all afternoon. Then we'll be finished for the year. Tonight everyone is invited to the harvest festival."

By late afternoon, Agassiz and the other workers had picked the last grapes. Harvest time was finished and now it was time for the celebration. Everyone hurried home to change into fresh clothes.

Louis's hands were purple with grape stains when he met Cecelia coming from the house. She had braided her hair and changed into her best clothing. Louis decided she looked beautiful.

Cecelia frowned when she saw her brother "Aren't you coming to the festival?"

"Not right away," he said. "Father must first calculate the profits of the harvest. If he made enough money, I can go to the university. If he didn't, I'll have to find a job."

Out on the hillside sounds of a songfest with yodeling began. Cecelia skipped away to join the fun. Louis walked slowly toward the house. Above the door, he read words engraved in wood: "What God hath founded builds on firm ground."

Louis went to his room. Objects of all kinds lined its walls; an insect collection, a flower pressed under glass, and shells arranged according to size. On a shelf above his desk, he had placed the blue vase from the lake. Louis looked at his greatest prize, a rock embedded with the outline of a fossil fish. Louis traced his fingers over the skeleton. How many years ago had the fish lived? What caused its body to be pressed into the stone? Above the fossil, he hung a colored drawing that represented his idea of what the fish had looked like when it was alive in an ancient sea.

Louis had learned so much already about birds, fish, snakes, frogs, and flowers. He felt he had only gained a glimpse into the wonders of God's creation. If only he could study science at the university!

He brushed his hair, put on a tie, and dressed for the festival, then walked downstairs to the kitchen. On the table, his father had counted out money in bills and coins and entered the total in ledger books.

"God has blessed us," Father Agassiz said with satisfaction. "We've had an abundant harvest." "How abundant?" Louis asked cautiously. Father Agassiz spoke gently. "My son, I know you hope to go to the university this fall."

Louis said quickly, "Nothing would please me more."

Father Agassiz continued. "The girls are older now. They need clothing suitable for young ladies. New dresses cost money." He looked at Louis. "It would be unfair to ask them to sacrifice so you could study further."

Louis protested. "I'm not asking for anyone to sacrifice. I only want freedom to pursue my own dreams."

Father Agassiz sounded impatient, "Your dreams require money. Uncle Mayor has agreed to take you into his business in Neuchatel as a bookkeeper."

Louis was outraged. "A bookkeeper? I want to study nature, not books." The thought of sitting all day at a desk and adding long rows of numbers was intolerable.

Father Agassiz looked stern. "You were not born to be a naturalist. Only the rich can afford science. Who would pay a man to collect rocks and study the scales of fish?"

Louis Agassiz continued to plead with his father. "But my grades at the academy were excellent."

Father Agassiz nodded. "I have letters from your teachers at the academy."

Louis said desperately, "And?"

"They say fine things about you," his father said. "But the academy prepared you just as well to work as a bookkeeper. You'll lead a life far more secure financially than you would as a scientist."

Louis tried once more. "I believe God has given me the ability to study nature because that's what he wants me to do."

Father Agassiz closed his ledger and locked his cash box. The subject of Louis's future was closed.

Louis walked out of the kitchen, tears of disappointment shining in his eyes.

Father Agassiz called, "You can't live on dreams, no matter how brilliant you are!"

## Chapter 2: Follow a Dream

Louis left the house and slowly walked toward the festival. His feet felt like lead although his ears began to ring with the sound of folk music.

All the workers were dressed in their finest clothing, the men in white shirts and sleeveless red jackets with silver buttons. Their dark trousers were tucked into long white woolen stockings, and shining buckles latched their shoes.

The girls wore white blouses with laced, black velvet vests. Their hair was braided and topped with tiny white lace caps, and around their ankles swished long full skirts of striped wool.

Many people including the man he had fished with that morning greeted Louis, but Louis was unable to join the fun. He sat down, then stood up restlessly and went for a walk. He paused near a large boulder on the edge of the lake and stared at the water, so deep in thought he didn't hear Cecelia approach him.

She spoke softly not wishing to startle him. "What is bothering my brother on a night when everyone else is happy?"

Louis's voice was soft too, but heavy with sadness. "The book of nature has been closed to me." Then, with effort, he turned away from his thoughts to consider his pretty sister. "You are enjoying the party tonight? Take care with all those attentive young men, Don't break too many hearts tonight."

Cecelia smiled, concern for Louis still evident in her eyes even as a handsome young man came to take her back to the festival. Louis sat on the rock and continued to stare across the lake.

At last, the harvest festival ended. Father Agassiz stepped onto the back of a wagon where he could be seen by everyone and then called for attention. When everyone became silent the preacher thanked the workers for their help with the crops and asked for one last song. His deep rich voice led the music:

Free and forever free,
ha-la-le-a-ho, ha-la-le-a-ho.
Care and labor now are gone,
hol-di-ri-di-a, hol-di-a.

After the festival, the family walked arm-in-arm back to the house for devotions before bedtime. Inside, Father Agassiz asked for his Bible, and Mother hurried to fetch it. Now the preacher began to read. Louis stared straight ahead, his eyes fixed. He barely heard the words.

"I returned, and saw under the sun," read the father, "that the race is not to the swift, nor the battle to the strong." He looked at Louis, but Louis sat still as stone. He sighed and resumed reading, "Neither yet bread to the wise, nor yet riches to men of understanding, nor yet favor to men of skill; but time and chance happeneth to them all."

Then Father Agassiz closed the Bible and bowed his head. Everyone repeated with him, "Oh, loving Lord, watch over our herds. Make our alp green. Keep our cows safe from snow, hail, or falling rocks. Amen."

Suddenly, before anyone else could move, Cecelia stood up. She looked at Louis, forcing his eyes to meet hers. Then she spoke. "Augustus, Olympia, and I have decided to save enough money so Louis can attend the university this fall."

"I'll work the ledgers and earn a daily wage," Augustus said quickly.

Mother Agassiz supported him. "Augustus is nimble with figures."

The obvious solidarity of everyone in the family in support of Louis startled Father Agassiz. He examined the faces of his children carefully, looking for signs of rebellion. Finding nothing but pleas for understanding in their eyes, he sighed and turned to his wife. "Very well," he said slowly, "Louis may attend the University of Heidelberg."

The children cheered, but before they could rush over to congratulate Louis, Father Agassiz silenced them. "You may go to the university," he said, "but you must study to become a physician."

Louis accepted the qualification immediately.

His father pushed back his chair and left the room.

Augustus hugged Louis. "I know you don't want to become a doctor," he said. "But you should have time to study nature while you work on your medical degree."

Louis nodded. "There are more than eight hours in a day. While I'm studying death and disease I can also learn what Heidelberg has to offer about animals, plants, and fossils." Then Louis looked at the others with determination in his eyes. "Nothing will keep me from my studies."

The University of Heidelberg was one of the best schools in Europe. In Switzerland, Louis had spoken French and some German. In Heidelberg however, he had to use German almost entirely, risking the ridicule of other students who viewed Swiss students as simple farmers with foreign accents.

At first, Louis thought he would be lonely, but soon found an unexpected friend in his roommate, Alex Braun. Alex came from a wealthy German family and like Louis, loved to explore the world of nature. The fact that Louis had a Swiss accent didn't matter to Alex at all, and the two boys soon became close friends.

Nature classes at Heidelberg fascinated the boys; the instructors expected students to explore the countryside to gather plants and animals to study. Nothing could have pleased these students more.

Alex's family lived near enough to visit on weekends. "Come home with me," Alex said. "We'll gather specimens after you meet my family."

His invitation was quickly accepted, and that weekend the boys left for the Braun home. As they neared the estate, they noted the ruins of an old castle on a hill. At the base of the hill sat a girl, sketching the ruins on a drawing board.

Alex waved at the girl. "That's my sister," he said.

She wore expensive clothing and spoke in the refined manner of the German aristocracy. Her well-bred appearance contrasted sharply with the rough field clothing and outdoor hardiness of Louis Agassiz.

Alex introduced his friend. "This is Louis Agassiz from Switzerland who is rooming with me at the university."

Cecelia examined the strong, firm lines of Louis's face, and pushed aside her castle sketch. She positioned a new sheet of paper on her board and began to sketch Louis with clean, accurate strokes.

Louis said her name. "Cecelia. That's my sister's name too, but I never knew it was so beautiful."

"Louis can name every animal," Alex said. "He knows every bird by its call and can identify any fish that swims."

Cecelia sketched on. "That is quite a feat," she said.

Louis's words tumbled out in a rush. "I still have so much to learn. So far I have only been able to study parts of nature. God's creation isn't a jumble of mixed-up pieces. They fit together in a pattern. The job of a naturalist is to trace out that pattern designed by the creator.

"Do you intend to define that pattern?" Cecelia asked.

Louis nodded. "I'm a medical student to please my father. But in my spare time, I study nature."

Cecelia turned the drawing board to show Louis the finished sketch.

Louis examined the drawing with pleasure. "This is my first portrait," he said.

"You may have it," Cecelia said. "My father is a botanist. I draw for him."

Louis was reminded of something. "Would you have time to sketch something for me?"

"If you wish," Cecelia said.

Louis reached into his canvas specimen bag and pulled out a dead fish. He held it up in the air by its tail.

Cecelia pulled away from its strong odor. "It, uh, smells. What is so important about this dead fish?"

"See this top fin?" Louis said, pointing his finger to show her. "A fin like this is seldom found in fish today, yet it was common in ancient fish. I've seen fins like this in fossils."

Cecelia agreed to do the fish despite its odor, so Louis left it with her while he and Alex prepared to go on to the Braun home. "I'll see you when you finish the drawing," Louis promised.

Time passed quickly at Heidelberg. Many times the professors would take classes exploring the countryside. Each student was assigned a different job such as gathering plants or minerals. Some made detailed drawings of layers of rocks in cliffs. Always Louis climbed the highest, took the greatest risks, and worked the hardest.

One day he climbed up the side of a cliff, and, hanging by a toehold, worked loose a stone. He examined it with satisfaction and climbed down from the cliff.

"Another fossil fish," Louis cried. He carried the prize to his instructor. "Will you identify it?"

The instructor looked unimpressed. "Fossils are so common that it is impossible to classify them properly. Professor Cuvier in France has tried to do it, but even he has failed."

"Why are fossils so common?" Louis asked.

The instructor explained. "Professor Cuvier believes long ago a series of worldwide disasters came upon the earth. The last disaster was the great flood at the time of Noah, described in the Bible."

He shook his head in disapproval. "Cuvier is an old man. Today's scientists reject his notions about worldwide disasters."

"His idea seems reasonable to me," Louis said boldly. What else could account for so many fossils? Why was this fish found so high above sea level? Might not the idea of a great flood explain it?"

The teacher was unwilling to accept Louis's reasoning. "Great disasters such as worldwide floods are too horrible to think about. Scientists prefer to believe the earth's climate has not changed."

"Shouldn't facts determine what a scientist believes rather than what he prefers to believe?" Louis asked.

The instructor smiled patiently at the persistence of his young student. "What are the facts, young man?"

Louis explained. "In Switzerland, we have found giant boulders that we call foundlings, or homeless children. Some force carried them far from their natural place in the mountains to the meadows of our farms."

A German student called out, "The farm boy clings to old-fashioned notions. Like all the Swiss he is one hundred years behind us."

Louis stiffened, but Alex whispered softly, "Do not take him personally. He insults anyone who is Swiss."

Louis was furious, however. "I will not be treated this way by anyone."

"Be cautious," Alex said, holding his friend by the arm. "The German is an expert fencer. He's trying to lure you into a sword fight."

The German student continued to sneer at Louis. "If you think you have been insulted, you should seek satisfaction in a duel. I would be most happy to meet you in the fencing hall."

Louis shouted, "Don't come alone. Bring the whole German team!"

The German grinned. "No one would want to miss the pleasure of watching you get put in place. Remember, farm boy, we use rapiers, not pitchforks."

## Chapter 3: The Great Baron

The next morning Louis entered the fencing hall. Many Swiss students gathered around him with a boisterous cheer, but boos and hisses from the German club forced Louis's eyes to evaluate his opponents.

The first challenger was tall and slender, the German club's finest fencer.

Louis refused to be intimidated. Something flickered deep in his eyes as he faced the German and announced, "Bring him on. I'm ready."

Louis stepped forward into the on-guard position. The German hesitated, momentarily put off by Louis's fluid movement and obvious familiarity with a rapier. The German hadn't expected this farmer to know how to hold a sword, much less how to use it. Now it was too late.

The German attacked suddenly, thrusting his sword forward with lightning speed. Louis was ready. He met the German's attack firmly, parried the thrust, and with a strong lunge forced the man backwards. One clean move with his sword suddenly rendered his opponent defenseless.

The German conceded reluctantly. "Touché." He had been touched by Louis's weapon and thus had lost the bout.

Louis saluted the fencer with his sword and stepped back to the runway.

The Swiss students cheered.

The second member of the German club stepped forward. He looked stronger than Louis, but Louis's probing sword soon found him less agile. The Germans tried to wear Louis down with a long game of defense, but Louis did not tire. He matched the German thrust for thrust until he sensed the German's attention was wandering. Then Louis lunged forward. Another victory.

The German students were silent. All eyes in the club turned to the man who had insulted Louis the day before.

He met their eyes boldly, lifting his chin, and stepped forward to challenge Louis. "Let's see if the Swiss grape picker can stand the sight of blood," he sneered.

Despite his bravado and reputation as a formidable fencer, this German exhibited little skill. Foolishly, he exposed himself time and again, but to take advantage of his bad form Louis had to strike within reach of the man's blade.

It took a long time, but finally, Louis touched the German with his blade. The match was over. Louis stepped back. It was too soon. The German lunged forward. His sword slashed through Louis's shirt.

"The German looks for blood," Louis said grimly. His movements became guarded and cautious. Fencing without masks was a deadly sport. The thrusts and parries of razor-sharp rapiers were sure to do more than draw blood. The German wanted to do more than injure his reputation. He wanted to kill Louis!

But Louis would not injure his opponent. Everything in him rebelled against the thought of bloodshed. So how was he to bring this fencer to defeat without injuring him? Time and again Louis touched his opponent, but the German would not admit defeat.

Suddenly the German lunged forward with his sword. Louis parried the strike quickly, gave a sharp tap on his opponent's blade, and thrust his weapon forward. The German, moving back, tripped and fell on his back. Louis positioned the tip of his rapier on the German's chest. The Swiss club cheered.

Now another German student stepped onto the runway. He had only words for Louis. His sword was firm in its sheath. "Our friend spoke in jest yesterday," he said to Louis. "Accept our regrets for his poorly chosen words. The duel is over."

Louis raised his rapier in salute, lowered it, and stepped off the runway. All the students in the hall cheered.

From that day, Louis's reputation at Heidelberg grew. Soon he became known not only for his skill in sports, but also for his great knowledge of science. The week after the duel Louis and Alex traveled to the Braun home. Despite Louis's protests, Alex told Cecelia about the duel.

"At first I thought Louis would end up looking like Swiss cheese," he said to his sister. "But before it was over I wondered what became of all those German experts. Louis won't kill animals to study them, but Germans would be wise to respect his rapier!"

Louis visited the Braun home many times during the school year. Each time he found himself drawn to Cecelia. He was attracted to her sensitive features and friendly nature, but most of all to her deep Christian faith. Louis knew she would make a wonderful wife. He wondered if she knew how much he already loved her.

As the school year ended, Louis decided to ask Cecelia to marry him. Now he just had to find the right words to ask her.

One night he finally found himself alone with the girl. His first words were hardly what he had rehearsed for this important occasion. Instead, he found himself apologizing. "I have spoken to famous men many times," he said, "even the Baron von Humboldt. But no one has given me more cause for nervousness than you." He gulped and looked down. His face felt hot.

Cecelia asked gently, "Would you like a sip of water?" Her eyes brightened mischievously. Louis continued bravely, "I wish to ask you to marry me. You are so important to me that I cannot imagine a future without you."

Cecelia's grin faded. "My parents would never permit me to marry a man who has no visible means of support."

Louis was shattered. "I have no money, little education, and no home. My future is uncertain. Yet when I receive my medical degree I will be able to provide a home for you."

Cecelia's voice was gentle. "I accept your proposal of marriage, Louis. But we must not mention this to father until after you have your medical degree."

Louis reached for his canvas specimen bag, and Cecelia's eyes sparkled. She expected something small and beautiful—like an engagement ring.

Louis pulled out something small and beautiful to him. It was a rock. "Here is another specimen for you to draw," he said briskly. "It's a fossil fish that was sent to me from a museum in Munich. Would you sketch it for me before I send it back?"

Cecelia sighed. "Do you need a wife or an artist'?"

"Both," Louis replied without hesitation.

"At least fossils don't smell," Cecelia said. "Remember your manners when you meet Baron Humboldt for breakfast. Dress like a student of medicine and not like a fish peddler."

Baron Humboldt was a great scientist who had traveled all over the world. In his studies, he made many important discoveries, for instance, he had found that the waters of oceans flow in great rivers called currents. One of these rivers, Humboldt's Current, was named after him.

Louis met the baron at an expensive restaurant where they were escorted to a private dining room. Humboldt said to the waiter, "Soup and oysters for myself. Beef with special sauce for my friend. We will both have pheasant and, when we have finished, cake and custard."

Humboldt looked around him with pleasure. "This restaurant is quite an improvement on some places I've eaten at throughout the years. One time on an expedition in Venezuela, we had to eat catfish cooked in mud."

Louis made a face.

"Actually, they were quite good. The natives showed us how to bake them. They also told me the truth about electric eels. You would be wise to remember this as a scientist; people know most about things in their own habitat, far more than any scientist in a lecture hall."

The waiter arrived with the first course, and Louis immediately picked up his fork. As a student, he seldom had enough to eat, and this meal at 8:00 in the morning looked especially good.

Humboldt continued to talk. "On the way back from South America I visited the United States and stayed with Thomas Jefferson, its president. One morning I was eating breakfast when Jefferson suddenly burst into the room. He held up a newspaper clipping for me to inspect.

"Read that, friend," Jefferson said. The article was shocking. A reporter had written terrible things about the president. When I finished reading the clipping, I looked up to find Jefferson grinning at me.

"How can you smile at a thing like that?" I asked. "in France the reporter's head would be topped off."

"That is France," the president said. "This is the United States. I am written about in this way every day, yet I don't complain because it is good that my people do their own thinking. Take this clipping with you. Remember it as a symbol of this nation."

Louis finished his beef and struggled to spoon up every bit of sauce. He was still hungry, and could not stop himself from staring at the baron's untouched soup. The man smiled and nudged his bowl toward Louis. Louis dipped in his spoon without hesitation.

The baron said, "Unrestrained expression of ideas is the sign of a free people. The scientists of Europe could profit a great deal from an environment in which everyone's ideas could be openly stated. Louis, you must visit the United States."

"There are many places I would like to visit," Louis said, putting his spoon down carefully. "Right now, however, I haven't the money to even go to Paris to meet with Professor Cuvier."

"Cuvier is the greatest fossil scientist alive today," Humboldt replied. "He has even unearthed and identified an elephant in the soil of Paris. You must visit him. Surely you have earned money enough to travel from the book you wrote, haven't you?"

"Not enough," Louis said, shaking his head sadly. "Your books are read by many people, but the book on fish I wrote has not sold well. I haven't earned nearly enough to go to Paris. Since my father considers it a great evil to spend money I have not yet earned, I will have to wait to visit the professor."

The baron asked, "What does your father think of your book?"

"I wanted to keep the book secret," said Louis. "I have been fitting the study of fossils in between my study of medicine. I didn't want Father to know about this until after I became a doctor. Now the book has changed all that. Father wrote me a letter."

"He says my priorities are all wrong." Louis reached for the letter in his pocket and read from it, wincing at the severity of its words.

"If it is essential to your happiness that you locate glaciers in the north and south poles to inspect hairs on a mammoth," wrote the father, "or that you dry your shirt in the sun of the tropics, at least wait till your trunk is packed and your passports signed before you tell us about it. But above all, before you aspire to any travels beyond Heidelberg, attain your first, most important goal. Get your physician's diploma!"

Baron Humboldt stared at the empty dishes on the table. Never had he seen anyone eat so much food as this student, who now sat back against his chair. The baron sighed and spoke to the boy. "I must write a letter to your parents telling them about the many gifts of their son. I will also urge them to give you permission to visit Paris."

"A visit to Paris would take time from medical studies," Louis pointed out.

The baron would not be discouraged. "Move to Paris then. You can study medicine there as well as Heidelberg."

The waiter approached the table. "Do you gentlemen wish to order lunch?"

Louis and Humboldt looked at the clock. It was nearly noon. They had talked for over three hours. It was time to leave.

"We must talk again to sharpen our wits," the baron said. "I don't know what I have done for you, but I do know how much you have stimulated me. Knowing you is a great pleasure."

Louis said humbly, "I have learned more in the last three hours from you than from all my other years of study." He smiled gratefully.

The baron reached in his pocket for an envelope. "You must never sacrifice your education because you lack money," he said, handing the envelope to the boy. "Consider this a loan you need only to repay with diligent work." Then the tall, aristocratic baron walked away.

Louis opened the envelope. It was filled with money! In a shaky voice, he whispered, "I can go to Paris!"

## Chapter 4: Animals in the Ice

Louis quickly arranged at Heidelberg for his trip to Paris and began to pack. When he had stuffed all his clothes into a traveling bag and tucked his sketchbook under his arm, he stood still long enough to describe his visit with the baron to Alex, his roommate.

"How he questioned me," he said. "And how much I learned from him during that short time; how to work, how to use my time, and what to study. Best of all, he even gave me money to go to Paris to meet Cuvier."

"Professor Cuvier may resent your intrusion into his field of fossil research," Alex warned. "He may not welcome the competition. I understand he is reputed to be cold and unfriendly."

"I have never tried to compete with Cuvier," Louis said. "In fact, I even dedicated my book to him. Maybe that will soften his heart."

When he arrived in Paris, Louis immediately set off to find the professor's home. Still carrying his bags, he examined a piece of paper with an address written on it. He matched the written address with a door number. This was the place. Taking a deep breath he stepped forward and pulled the bell.

A butler opened the door. "Professor Cuvier is receiving no one," he said. "Leave your calling card, please. I will give it to him."

"I have no cards," said Louis.

"Good day," the butler said. He began to close the door.

Louis quickly handed him his sketchbook. "Give this to Professor Cuvier," he said.

"Very well," the butler said. "Wait here."

Louis waited anxiously. It would be awful to have come to Paris without ever having had the opportunity to meet the world-famous scientist.

Minutes passed, each one feeling like an hour to the young man. Finally, the door opened. The butler stood like a soldier, disapproval hard on his face. He spoke reluctantly. "Professor Cuvier is in the museum supervising work on a new display. I'll take you to him."

Inside the house, Louis followed the butler down a long hall leading directly to the Museum of Natural History, one of Paris's most famous museums.

Professor Cuvier was an old man with a weak voice, His thick hair was snow white, and he walked with a cane.

As Louis introduced himself, a spark of interest brightened the professor's eyes. The old man began to speak about his museum. As he talked about specimens and displays, he appeared years younger. He built many displays in the museum, featuring his most important discoveries.

Suddenly he stopped and thumped his cane on the marble floor. "Right here in Paris, I found the bones of giant lizards. I discovered fossils of other giant beasts as well; ancestors of hippopotamuses, rhinoceros, and bears. Many of these animals have vanished from the face of the earth. Why did they become extinct?"

Louis spoke, choosing his words carefully. "Somehow your question reminds me of the woolly mammoth found in Siberia."

Cuvier was delighted with Louis's comment. "Come, I have something for you to see." He led Louis to a display containing the bones and man-made restoration of a woolly mammoth. The giant beast looked something like an elephant, yet this elephant no longer lived anywhere on earth. Woolly mammoths were extinct.

"Fishermen discovered this one," the professor said. "Frozen in ice along the Obi River in Siberia. It appeared to have died only a few hours before discovery. Yet it had been dead for thousands of years!"

Louis touched a sample of the animal's skin. Twelve-inch hair like thick twine grew from the hide.

"I built this life-size model of what I thought the beast looked like," Cuvier said. "Isn't it strange such strong and powerful animals died in such great numbers that they disappeared totally from the face of the earth?"

"It doesn't seem possible," Louis said.

Cuvier went on. "Only a worldwide disaster could have caused this. From my observations and discoveries, I can only conclude that life on this earth has not been peaceful. In many layers of rock, I have found fossils which record tragedies that killed hundreds of thousands of animals."

"Do you speak of worldwide catastrophes such as the biblical flood?" Louis asked.

Cuvier nodded. "Every living thing was destroyed which was upon the face of the ground," he said, quoting from the Bible, "both man, and cattle, and the creeping things, and the fowl of the heaven; and they were destroyed from the earth."

Louis said carefully, "Many scientists don't believe that flood actually happened."

Cuvier looked troubled. "I am aware of that. Some scientists call the flood superstitious nonsense." He looked at Louis defiantly. "They have not examined the evidence the way I have. Scientists who believe the earth has not undergone great change take the easy path. It makes their job easier. They are not only lazy, they are also wrong!" Suddenly the old man grew calm. "Some day, someone will prove them wrong. I wonder, will it be You, Mr. Agassiz?"

It was time for Louis to leave. The old professor seemed to be very frail after his lengthy discussion with the student. Yet he reminded Louis of church services the next day and invited him to spend Sunday with him.

The next morning Louis and Professor Cuvier walked out of church after the service together. The day was bright and sunny, and the professor stopped on the steps to enjoy the warm weather.

Suddenly he turned to Louis and asked, "Will you work with me on fossil fish studies?"

Louis's mouth went dry. "I would count it the highest honor," he said gratefully, "to be considered your assistant."

Cuvier shook his head. "No. It's too late for that. Maybe a year ago you might have been my assistant, but not today. I am too old and frail."

Louis bit his lip. "Then how could I help you?"

Cuvier explained. "I will give you all of my notes and put you in charge of all my collections. You can work in my house, but we'll not be working together. The work is all yours."

Louis found it difficult to speak. "The specimens are priceless, the most wonderful gift I could ever receive. I will work hard to prove myself worthy of your trust."

Louis Agassiz spent a year in Paris. During that time he studied books on medicine and attended lectures by medical experts. He also measured, categorized, and organized fossil records in Professor Cuvier's museum. He visited every other natural history museum in Paris.

Slowly his work on fossil fishes took shape. It would be one of the largest and most detailed works on ancient life ever produced. The ancient seas in which they swam came alive to Louis, but work on fish took much of his time. He was unable to visit any of the famous sights in Paris.

One day Professor Cuvier visited the workshop. He seemed frailer than usual, even holding onto the doorknob for support. Louis was leaning over a worktable, measuring bones. He sketched them, then wrote a detailed description of each in weight, size, and color.

At first Louis didn't even notice the professor. When he finally did, he was profuse with apology. Louis explained why he worked so intently. "Unless I appear promptly at

5:00 in my boarding house, I will miss dinner. I will return for more study after dinner."

"Your enthusiasm is good," said Cuvier, "but too much work can kill a person. You have yet to see the sights of Paris; its art museums, Notre Dame, and the opera. Consider this, son. Chopin is giving a piano recital this very evening!"

Louis shook his head. "I have little time to even visit the sea coast, much less the sights of Paris."

Professor Cuvier walked out of the museum, and Louis continued to work as if he had not been interrupted. He was totally unaware of how sick the professor looked.

When Louis came to the workshop the next day, the professor's doctor met him at the door.

"Who is ill?" Louis asked.

"Professor Cuvier," the doctor said.

"How ill is he?" asked Louis. He began to realize something was terribly wrong.

"Professor Cuvier died during the night," the doctor said. "It was a stroke. I told him for years to take time off from work. He finally took my advice and slowed down, but it was too late."

The doctor looked closely at Louis. "Some day you must learn this; the human body can take only so much. You must take time to relax. You would do well, my son, to vary the pace of your studies. Take some time off."

After the death of Cuvier, Louis did take a brief vacation when Alex Braun came to Paris. Together they toured the city.

"Let's go to the sea coast at Normandy," Louis suggested. "We can stroll the beach there."

"Normandy is one hundred miles away!" Alex said in astonishment.

"Walking is good exercise," the science student said with a grin.

Louis was delighted with the ceaseless sound of waves against the seashore at Normandy. He was fascinated with advancing and receding tides, and everywhere he delighted in new species of birds and seashells.

"At last I have looked upon the sea and its riches," Louis murmured softly. "I had almost despaired of seeing it."

He walked along the beach with Alex and drew in a breath of salt air. "Where is America?" Louis asked.

Alex pointed across the water. "That way, about four thousand miles."

"There is much over there that even Baron Humboldt missed," Louis said. "He never climbed the Rockies or sailed the Great Lakes. Someday I will see them for myself."

At night the young men built a fire on the beach. As the sky began to glitter with thousands of stars Louis asked quietly, "How is Cecelia?"

"Cecelia sends you greetings," Alex said. He hesitated before speaking again, then continued. "Cecelia cannot wait to marry you forever. Father has arranged a marriage."

Louis interrupted, "For Cecelia?" His heart sank.

"No," said Alex quickly. "For Emma, our older sister. Emma was supposed to marry Karl Schimper. He couldn't find a good job, however, so father arranged for her to marry a wealthy, middle-aged man."

Alex added gently, "The same thing could happen to Cecelia and she would be heartbroken. Cecelia might be willing to wait forever for You, Louis, but Father will not. I suggest you marry her without delay."

"I cannot marry her until I have a medical degree," Louis said, "Without that I would be unable to support her."

"What more must you do?"

"I have finished all my work in Paris," Louis said. "My first book of fossil fish studies is at the printer. I have also completed my medical studies. It is time for me to return to Heidelberg."

"What then?" Alex asked.

"The medical board of Heidelberg will examine my knowledge of medicine, then decide whether I should receive the degree to practice medicine. My future, and Cecelia's, is in their hands."

## Chapter 5: Doctor Louis

Louis waited outside the boardroom in Heidelberg. Dark-suited doctors and medical instructors filed past him into the meeting room. The last one stopped in front of Louis, raising his spectacles to examine the young man. Then he pointed to a wooden bench by the door. "You must wait here until we reach our decision."

Louis sat down, his knees weak.

The man rested his hand on Louis's shoulder. "We have already read your written report. After we have reached a decision on that paper we will call you in to ask you some questions."

"What will you do if my written report isn't good enough?" Louis asked.

"There would be little need for us to question you further," the examiner said. "You would have to try again a year later."

Louis waited in the gloomy hall, miserably uncomfortable on his wooden bench. Through heavy doors, he could hear muffled voices in prolonged discussion, but no matter how hard he listened, he could not hear what was said.

Finally, a man opened the door and asked Louis to come inside. Louis stepped into the room and tried to look beyond the many eyes focused on him as he walked to a long table.

The dean spoke first. "We have examined your paper, Louis Agassiz," he said gravely. "You discussed in it seventy-six problems of medicine. After much discussion, we have decided there is no need to question you. "

Louis was crushed. When he could finally speak, he said in a low voice, "I am sorry you didn't find the work acceptable." He turned away toward the door.

The dean quickly restrained him. "You don't understand," he said. "Your work is so well written that it is enough. Questions and answers aren't necessary. We are most honored to grant you a medical diploma. Already you have earned a great reputation in Europe. Congratulations, Doctor Agassiz."

The next day Louis took Cecelia for a horse and buggy ride in the country. He had so much to tell her! "The city of Neuchatel is opening a museum and a high school," he said enthusiastically. "I have been invited to teach there by Monsieur de Coulon. He expects at least one hundred students to enroll."

Cecelia was less enthusiastic, "Neuchatel is a tiny Swiss town, hardly an important scientific post. Will you be happy there?"

"No place could be better," Louis said. "Nearby is a lake, meadows, wooded hills. So are the glaciers 1 have always wanted to study. But first I must ask you something very important."

Cecelia held her breath.

"When can we get married?"

Cecelia sighed. "I thought you'd never ask."

After the wedding in Germany, Louis and Cecelia returned to Switzerland. Cecelia made use of her skills by taking charge of finding a place to live and outfitting it with furniture.

It wasn't long before Louis was able to lead his first expedition to a glacier. After a long trek up a mountain, the explorers found themselves gazing, awestruck, at a long expanse of ice. At the base of the glacier water from melting ice had created a huge waterfall. The sound was deafening as water poured out from under the ice.

The guide yelled over the noise. "The cave under the glacier collects water which drains into it through cracks called crevasses in the ice. It comes out here."

Louis walked closer to the glacier. The wind swirled around him. The groaning and cracking of ice was so loud he couldn't hear himself speak. Gusts of air shot out of the cave, whipping up his coattails.

"The glacier is like a living thing," Louis said. He reached down for a handful of loose soil, which he identified as rocks that had been ground down by the ice into a fine powder.

The explorers scrambled over a ridge of gravel and boulders and climbed onto the ice. Soon they came to a crevasse. Louis walked to the edge. Water swirled deep down in the crack. Suddenly Louis felt dizzy. What would happen if someone fell into this bottomless hole, he wondered.

"How thick is the ice here?" he asked the guide. "No one knows," the man said. "But this crevasse appears to be about two hundred feet deep. Many cracks are even deeper."

They climbed higher onto the ice. Then Louis conducted a strange experiment. He asked the explorers to drive ten-foot-long stakes deep into the ice. The markers were positioned in a straight line across the glacier, only their tops showing over the ice.

Louis explained his reason for the markers to the guide. "Many local people claim glaciers move from year to year. I want to verify that observation, even though most scientists say it is impossible."

The guide shrugged. "I've heard people talk about ice that travels, but I've never seen a glacier move."

Louis went on patiently. "Movement might be so slow that it can't be seen by the naked eye. These stakes should settle the matter. I'll return next year, and if this glacier moves it will carry the markers down into the valley. "

On the way back to Neuchatel Louis and the men stopped to rest by an enormous boulder. The rock, which was almost as large as a house, stood alone in a meadow. Louis peered at the rock. "The stone is granite, but not at all like the other stones found in this area."

"How did it get here?" the guide asked.

"I believe a glacier pushed it here." Louis pointed to a low mound of dirt. "Glaciers were once tremendous rivers of ice. They filled the valleys to the brim, poured down mountains, and spread over the plains. Then they melted, leaving small hills of stone and rock."

It was time to leave. Louis left the glacier reluctantly and turned toward home and Cecelia. Their little apartment in Neuchatel was already crammed with scientific samples, and open boxes and packing crates had yet to be unpacked.

"Monsieur de Coulon sent a message," Cecelia said. "He is on his way over to meet us."

Monsieur de Coulon was at the door in minutes. He was a small, bright-eyed man whose friendly smile never left his lips. He immediately embraced Louis and cried, "Welcome to Neuchatel, Dr. Agassiz!"

"I am ready to go to work," Louis said. "Where is the museum?"

Monsieur de Coulon looked down. He spoke apologetically. "The town doesn't have a museum yet. We want you to start one."

Louis was visibly upset. "I was told the town already had a museum. "You have a budding for it, don't you?"

Monsieur de Coulon shook his head.

"A laboratory?" Louis asked.

Again the man shook his head.

"Equipment?" Louis asked.

"No." The sound was barely audible.

"Classroom?"

A smile suddenly lit up the little man's face. "Oh yes, monsieur, we have a classroom for you."

Louis looked at the man suspiciously. "Where?"

"Ah." Monsieur de Coulon hesitated, then spoke again. "The classroom is in the town hall. I have already ordered a blackboard for your lecture tonight."

"Tonight?"

"But of course," the little man said. "The people of Neuchatel expect it. This is a humble audience, Dr. Agassiz. Many have no use for books. They won't like you if you are superior to them. I hope you will speak to them about things they will be interested in." The Frenchman bowed deeply and left.

Louis's mouth hung open.

Cecelia gently pressed it shut. "That sounded like a threat," she observed.

Louis nodded. "I think it was. If I do not impress the people tonight, I may no longer have a job."

Cecelia looked worried. "What will you lecture about?"

Louis looked around at all his specimens. "I don't know yet. Would you please locate your sketch of the butterfly emerging from its cocoon?"

"And where is my microscope?"

Chapter 6: Rivers of Ice

That night people crowded into the town hall. Everywhere were little groups of people from the country and city, chattering and twisting on hard-backed chairs. Louis peered cautiously at his audience and quickly decided it would be difficult to please.

But he had little time to worry. Monsieur de Coulon was already at the lectern, introducing the young doctor. Louis swallowed hard and stepped onto the platform. He looked to his left at a stand featuring Cecelia's butterfly sketch, which had been covered with a black cloth. On a table next to the stand were positioned Louis's microscope and a small assortment of shells and fossils. Behind all this was the new blackboard.

The people stopped talking and stared at the young man.

Louis took a deep breath and began to talk. "Some people say studying nature is a waste of time. We will never learn all its secrets, they say. They believe the earth is controlled by blind forces that act by chance." Louis paused to look at the audience. "I believe everything I see proclaims aloud the God I love. Scientists did not invent the system of order in nature. Instead, they have merely identified what was placed there at the dawn of creation by the Almighty."

This afternoon as I walked to the lecture hall a hawk swooped over my head. He had sharp claws like this.

Louis sketched claws on the blackboard.

You would not expect a hawk to have hooves or a cow to have sharp teeth. Why? The hawk's beak and claws are adapted to catch and hold the field mice which are its food. The cow, on the other hand, has flat teeth used for grinding hay so it can be digested. The plan of God matches the animals with their way of living.

Louis looked at the people. He was gratified to see he had their attention.

He went on. "Professor Cuvier was the first scientist to write about this structural order. He said the teeth and claws of an animal can be used to identify what kind of food it eats." Louis decided to tell a story to illustrate his point. "One night one of Cuvier's students dressed up in a devil's costume. He and his friends sneaked into the professor's room and one of them whispered 'Cuvier, wake up. I have come to eat you!' You know what Cuvier did? He opened one eye, looked at the devil's horns, and said, 'Creatures with horns and hooves eat grass. You can't eat me.' Then he went back to sleep."

There were sudden laughs from Louis's audience.

"A scientist's job is to observe, not talk," Louis said. He pulled off the cloth covering Cecelia's drawing, and people gasped at the beauty of the butterfly.

"Anyone who wishes to may come forward to examine the displays. I will also try to answer any questions you may have. Or if you wish, you may leave. The lecture is over."

No one left. Instead, several men rose to question Louis.

"Why is science so important?" a farmer asked. "How can it help us?"

"What is a newborn baby good for?" Louis replied. "We don't know what today's discoveries will lead to yet."

Louis looked at a businessman. "Science may someday help the watches you make keep better time."

Louis turned to the farmer, "It may help you age cheese faster, grow grain in cooler climates, or even produce better cows and sheep."

"Welcome to Neuchatel," said the farmer with a smile and a handshake.

Suddenly several children ran forward to ask questions. One boy looked at Louis in wonder. "Do you know everything?" he asked.

Louis answered with words from the Bible: "Knowest thou the ordinances of heaven? Canst thou set the dominion thereof in the earth?" Then he went on to explain. He needed to investigate so many areas. For example, turtles exist in abundance in this area of Switzerland, but has anyone ever wondered how long it takes turtle eggs to hatch?

The boy looked confused. "'Turtle eggs? They aren't important.

"The study of nature often focuses on seemingly unimportant things," Louis said sternly. "Each scientist must learn to ask the right questions. For instance, when I look at glaciers I ask, what gave them birth? How long ago did that happen? Have glaciers always stayed the same or do they change? Those are questions I hope to answer someday."

Finally, people stopped asking questions, and the hall emptied. A janitor began sweeping the floor, putting chairs back in place. Louis and Cecelia packed the exhibit.

Now Monsieur de Coulon stepped forward to shake Louis's hand. "A most successful evening," he said, beaming. "You are to be commended, Doctor."

"My next audience may not prove to be as easy to please," Louis said. "Important scientists from Europe, members of the Helvetic Association, will come to hear me speak."

"What will you speak to them about?" asked the little man. "Fossil fish?"

"They will expect me to lecture on that subject," Louis replied thoughtfully. "It is the subject I know best."

Cecelia added, "The Geological Society of London awarded him a medal for his book on fossil fish."

"Cecelia's splendid drawings deserve credit for the success of my book among scientists," Louis said. "Still, I'll not be addressing the Helvetic Association on the subject of fossil fish."

"What will you talk about?" de Coulon asked.

"I'll tell them about the ice age," Louis said. He looked serious. "I don't expect them to agree with my views at all."

Later that month Louis Agassiz did play host to the grand old men of science. These were world-famous men, icy and aloof in their expertise. They had traveled from Paris, Berlin, Frankfurt, and London. Even Baron Humboldt had come.

Louis had tacked a map of the world to the wall. On it he had colored in the area he believed had been affected by the ice age. When the men realized what Louis was going to speak about, their faces tightened with disapproval.

Even after he began to speak, their faces did not soften. Time after time he saw them shake their heads in anger.

Louis finally came to the end of his speech. "I will conclude with this observation: Many years ago a long winter settled over a land previously covered with rich vegetation, where great beasts like those found in India and Africa freely roamed."

The scientists were restless now. Some pushed back their chairs and whispered to each other. Others shook their heads. A German in the front row spoke to Baron Humboldt in a loud voice. "Agassiz is not a geologist, he's a zoologist. What does he know about glaciers?"

Louis Agassiz had crossed the line between scientific disciplines. In those days chemists were not to study medicine; people who studied rocks and minerals were not to bother with living things such as plants and animals. Louis Agassiz had dared to combine biology with geology; he had studied both living things and the history of the earth.

"Preposterous!" said an English doctor. "Agassiz should stick with fossil fish."

Louis tried to finish his speech. "Death entered with its terrors. With one blow of its violent hand, it destroyed a mighty creation and wrapped all nature in a shroud of ice."

The audience booed. Baron Humboldt looked behind him, distressed by the conduct of his friends. Yet even he didn't seem to be happy with Louis's speech. A Frenchman tapped the floor with his cane, saying straight out, "It is silly to think glaciers covered the whole world."

The German nodded his head. "Moving glaciers? Never! Ice stays in the Alps. It does not travel anywhere else."

Louis said, "During the ice age glaciers did not stay on mountains. They grew until they covered the entire northern hemisphere. How else can we explain finding woolly mammoths still buried in ice?"

"Where is your proof?" a scientist cried.

"Observe the world around you. The proof is all over Europe," Louis said calmly. "Smooth round hills show the marks of incredibly heavy ice which once pressed down on them. Fields are littered with boulders and piles of stones which were left by glaciers."

Louis unwrapped a flat rock that was marked with deep scratches. "Here's more proof," he said. "Look at these scratches. Once a glacier moved over this stone and created the marks."

"Those scratches are nothing but sled marks," a man said from the back of the room.

Everyone laughed. Louis realized no one believed his theory. People started to rise from their chairs and head for the door. Soon the room was nearly empty. Louis looked around him wearily and met the angry eyes of Alexander von Humboldt.

"How could they insult the great Louis Agassiz? Unthinkable." The baron gently put his hand on the young man's arm and murmured, "Protect your fine reputation. Leave glaciers alone."

"Have you forgotten President Jefferson's newspaper clipping?" Louis asked with flashing eyes. "Is not the free exchange of ideas the sign of a healthy scientific community?"

"The ice age theory appeals only to the popular imagination," the baron said. "Many scientists are jealous of you, resenting the attention you get in the press. Other scientists think you are trying to make people believe your theory is established fact."

"What can I do to change their minds?"

Humboldt sighed. "You will erase their skepticism only if you can show them exactly how grooves were cut into stone. Then you must prove that glaciers do indeed move!"

Louis nodded. "You are right. I must do more study and gather more evidence. I have still only found pieces of the puzzle. I haven't yet put them together. But I will!"

Chapter 7: Hotel on a Glacier

In time Louis Agassiz wrote several more successful books. With some help from the city of Neuchatel, he bought the entire apartment building in which he lived, hoping to convert part of it into a museum.

Now that hope had finally become a reality. The workmen were already knocking down walls to create space for the museum. In a nearby building, printers were installing a printing press.

The printer adjusted the press, then fed it a sheet of paper. His assistant removed the first sheet and brought it to Louis. Louis peered at it, shook his head, and waved it away.

Cecelia brought him a tray of food. Off to the side of the food she had placed a stack of letters.

"Your mail is getting out of hand," Cecelia said.

"They're still complaining about my ice age lecture," Louis said. "Baron Humboldt is right. Scientific facts must support theory. I must strengthen my argument. I must find more proof that glaciers move. I'll go back to the Aar glacier to study it properly."

"When?" Cecelia asked. "In between your teaching, speech-making, book writing, and museum fundraising? Even in the summer, you are too busy tramping all over with that terrible Armand Gressly looking for fossils."

Louis defended his friend. "Armand Gressly may be untidy in dress, but he is a brilliant assistant. Nothing escapes his eyes. He is one person I must take back to the glacier. I will need a fully equipped expedition to penetrate the heart of the glacier."

Cecelia was a fragile woman who found it painful to cope with the controversy Louis generated. She was uneasy with his restless activity at Neuchatel too, preferring a settled routine with time off for relaxation.

Another wall crashed down as workmen pounded, and Cecelia felt weak. Louis looked at her ashen face with concern. "Are you all right?" he asked. "You are such a help to me with my work. I don't know what I'd do without you."

"I'll not be able to help much longer," she said shyly.

Louis felt panicky. "Are you ill?"

Cecelia smiled. "Not unless motherhood is a disease. I am going to have a baby in a few months."

Louis was speechless. A baby! He was quiet so long his wife almost became alarmed. Finally, he managed to gasp, "It's wonderful."

"But I doubt I will have time left to sketch or answer letters," Cecelia said.

"We must make other arrangements then," Louis replied quickly. "A child! Of course, you must leave paints and pencils to raise the child. Still -- I must return to the glacier before you give birth. Now who will accompany me?"

In time Louis decided to take five men. They met at his home, prepared for the climb with heavy woolen clothing and spiked boots. In the lead was to be Jacob Leuthold, a professional guide. Louis Agassiz would follow him, then Armand Gressly, his field assistant. Next was Francois de Pourtales, a student, followed by Jean Wahren, a porter who resented not being used as head guide. The last person in the party was Professor Burkhardt.

As Louis looked at each man now, he remembered how he had explained to Cecelia why he had chosen them. "Jacob Leuthold is the best professional guide in the Alps," he said. "Armand Gressly will collect fossils. Francois de Pourtales has always taken top honors in my classes and is vitally interested in zoology. I'll put him in charge of collecting plants and describing animals. Jean Wahren is a good porter, although he thinks he should be the head guide. Professor Burkhardt is an excellent weather forecaster, an absolute necessity for mountain climbs. He's also a good artist."

Several days later the men worked their way up the enormous glacier. One of them pointed to a pile of broken timbers that looked like a building that had been smashed. They climbed another mile before resting.

Jacob consulted his map. "About twelve years ago a monk named Hugi built a shack on the glacier. His cabin might make a good base of operations."

"Where is the cabin supposed to be?" Louis asked.

"According to the map, right there."

Louis scanned the ice. "There is no cabin."

Francois the young student spoke. "What about the broken lumber a mile from here? Could that be the remains of the cabin?"

"It's too far away," Jacob asserted.

Suddenly Louis became excited. "Don't you see what has happened?" he said. "The cabin was carried down the glacier by moving ice. In twelve years this glacier has moved five thousand feet!"

The porter grew restless. "It is dangerous to be on the ice after dark. We must find a place to camp."

Louis objected. "We can hardly learn the secrets of the ice by living away from it. We must search for a sheltered spot in which to make camp."

The men separated into pairs, searching for a place. It was after the sun had begun to sink and long shadows had darkened the ice that Jacob and the professor raised a flag, signaling they had found a spot. The group of tired men looked at a huge slate boulder half buried in the ice.

Jacob pointed to a slab jutting over at the top. "At least we'll have a roof over our heads."

"There are enough small stones around to build walls on two sides," Louis said. "We can sling a blanket across the front. This will do nicely."

"A regular hotel," Francois added with a smile. Louis grinned. "We'll call it the Hotel des Neuchatelois."

After everything was in place, Louis wrote Hotel de Neuchatelois on a piece of wood and hung it from the roof.

Then each of the explorers scratched their names into the slate ceiling. They lit a fire and settled down for the night.

Louis awoke the next morning to find that fog had settled over the glacier. The camp appeared to be floating on a cloud. Louis jabbed the professor with his toe. "You didn't predict this," he said unhappily.

"It will burn away later in the day," Burkhardt said. He looked around. "The light isn't right for drawing, anyway." He turned over and went back to sleep.

Louis shook him awake. "I need to locate several large boulders. We must mark their positions on the glacier in relation to the valley walls. Later we'll compare those positions and see if the stones have moved."

The mist soon dissolved, and the explorers climbed high on the ice. Along the way, they stopped to take measurements. Some drew diagrams while others collected rocks that had been polished smooth by the ice. Several times they stopped to cut samples out of the ice.

As the team moved higher they were forced to cross over deep cracks called crevasses. Jacob led the way, testing the ice. Once he pointed to a clear section of ice. A blue glow shone from an opening in the ice.

"Watch out," Jacob said. "Blue ice is dangerous. Underneath it is a crevasse. One false move and you would plunge deep into a crack."

Soon the sun came out, warming the mountain air. The men removed their heavy jackets as they came to a spot where the glacier squeezed through a mountain pass.

Jacob looked at the towering mountains on each side and looked worried. "When the weather warms, boulders work loose," he said. "Watch out for failing rock. Even the smallest pebbles can start an avalanche."

Only a few moments later the party had to scramble for safety as a small avalanche started by a tumbling rock barely missed them.

Louis asked if anyone was hurt. Before they could answer a sound like a rifle shot echoed across the glacier.

The ice heaved and a sharp, cracking sound shook the glacier like an earthquake.

"What's happening?" Francois asked.

Jacob said, "The ice is breaking!"

Rumbling and cracking sounds became louder until the glacier groaned and a crevasse opened. Part of the snow and rock from the avalanche dropped deep into the crack.

Francois was so close to the crack that when things began to slide, he found himself also slipping toward it. Only Louis's quick thinking and fast action saved him from sure death.

Jacob quickly led everyone away from the crevasse.

Louis was silent for a long time. Then he spoke.

"The ice must be under strain where it squeezes through the pass. When that falling rock hit, it must have created such additional pressure that the glacier had to crack."

"I would prefer to learn about the birth of crevasses under somewhat safer conditions," said Francois who still looked shaken.

Louis walked to the side of the glacier where it pressed against the mountain wall. He pushed his fingers into deep grooves in the rock. Nearby a boulder was wedged between the glacier and the wall.

"This is why there are grooves in the stone," he said softly.

He called the others over to see the evidence.

"The glacier pressed that rock against the valley wall," Louis said. "The pressure of the moving ice was so great it cut deep scratches in the rock."

Now Louis pointed ahead. "Last year Jacob and I drove stakes into the ice. They should be just ahead. We'll see if the glacier has moved them."

Armand Gressly found the first stake that was lying in a puddle of water. Sadly, he carried it over to Louis, who had spotted another stake. It too was surrounded by water.

"The ice melted," said Greesly.

"We didn't learn how fast the ice moves," Louis said. "But we did learn how fast it melts! We'll have to use longer stakes."

The men began to drill holes for new stakes. The new poles would be eighteen feet long, and three times as tall as a man.

Again the poles were set in a straight line across the width of the glacier so that later Louis could measure how far the ice had moved.

As weeks passed the hotel became too small for the men and their equipment. They decided to build a log cabin next to the old campsite. Then they sent for more supplies.

The professor was working on paintings of the glacier and also the towering Jungfrau. One evening he sighed and put down his brush. "It is time to leave camp and go home," he said. "Winter will soon be here. We cannot survive the storms on the ice."

"I have one more experiment," said Louis.

"What is that?" Francois asked.

"I must measure the thickness of the ice," Louis said. "I ordered a supply of metal rods to be brought up from the valley. We'll pound them into the ice until we reach stone."

Jean Wahren and Armand Gressly hammered the rods into the ice while Louis kept track of how many were used. Finally, Wahren handed his partner the last rod. "There are no more," he said.

"How deep have we hammered?" asked Gressly.

Louis multiplied the number of rods by their length. "According to my figures about six hundred feet. Yet we haven't reached the bottom. Pull out the rods. We'll bring more of them next year."

The men packed their equipment and stored it in the hotel. Then they proceeded down the glacier.

Louis Agassiz's first child was born soon after he returned from the glacier. "A son," Cecelia said, as she smiled weakly at the tiny infant.

Louis grinned with obvious and honest pride. "What shall we name him?"

"Alex," Cecelia said immediately, "in honor of your friend."

"Alex is your brother's name too," Louis teased.

Cecelia smiled. "So we both like the name. Are you done with the glacier?"

"For this year," Louis said, "But next spring we must return to finish our work."

"What needs to be done?" Cecelia asked.

"We must drive more rods into the ice to determine its thickness. And I still want to lower scientific instruments into a well."

"What is a well?" Cecelia asked, tracing her finger lightly over the baby's cheek.

"At various places on the glacier," Louis said, "water melts into the ice. Eventually, it makes such deep holes that they penetrate into the heart of the glacier."

"You must be careful," Cecelia said with a shudder.

"Of course," Louis promised. His thoughts were far away on a fog-shrouded mountaintop where ice groaned and cracked.

Chapter 8: The Heart of the Glacier

The next year Louis returned to the glacier with his assistants. One of their first tasks was measuring the glacier, so it wasn't long before Gressly was pounding metal rods into the ice. Rod after rod disappeared until finally the ice would hold no more.

"That's it," Armand said triumphantly, "We've struck bedrock. "

"One thousand feet!" Louis said. "Imagine, nearly a quarter mile of ice."

Jacob, Wahren, and Francois now searched the glacier for a well. When they had located one, they positioned a huge tripod over it. The professor sat on the ice and sketched the busy men.

First, they tied ropes to the tripod, then fastened the other ends to a board which created a swing on which Louis Agassiz would sit as he was lowered into the well.

"Wells are so common," Louis said. "I can't resist trying to investigate them."

"They are common, to be sure," Jacob said. "But nobody has yet tried to enter those dark holes."

"It's time someone did," Louis said impatiently. He stepped over to the swing.

One rope would lower Louis into the well. Another one would be used for signaling. Louis explained the rope signals to Francois. "Jacob will lower me at a steady speed. If I want to stop, I will pull the hand rope once. If I pull twice, lower me further. If I pull hard jerk me up at once; it will be a sign I am in danger."

Francois looked worried. "If the ice moves as you say, aren't you afraid the well might close?"

"I won't be down that long," Louis said.

The boy was persistent. "Why don't we just lower instruments?"

"I must find out what it is like inside the glacier," Louis said. "I can see much that instruments might miss. "

Gressly smiled. "In that case, shouldn't we lower the professor so he could make a drawing?"

"I'll stick with painting the outside of the glacier," Burkhardt said slowly. "Louis can explore the inside."

Louis climbed into the sling. "Lower away!" he called.

Francois lay flat on the ice at the edge of the well. From there, he played out the signal cord. Jacob and Armand slowly lowered Louis into the well.

Sunlight filtered through the ice as Louis slowly descended into the well. There was an eerie blue glow all around him. He looked closely at the ice and marveled at its beautiful blue-white layers. Then he pulled on the rope to signal a stop. Francois noted the movement with a twinge of fear.

"The ice is in layers," yelled Louis, "like the growth rings of a tree."

He pulled twice on the signal rope. "Lower away!"

It was darker as Louis swung deeper. Then suddenly he saw beneath him the well was dividing. He pulled on the rope.

"What's happening?" Armand demanded as he held the rope steady.

"He says he's reached a place where the well divides," Francois said. "How far down is he?"

Jacob examined the coil of unused rope. "About eighty feet."

The rope jerked twice.

"Lower away," Francois said.

Louis passed through part of the well where huge icicles hung. Some were almost thirty-six inches long, tapering from bases four or five inches wide to sharp points. Louis wondered what would happen to him if one of the ice fingers broke loose and fell on him.

Now the icicles disappeared. Deeper into the well the walls were smooth, glistening with water. Louis ran his fingers along the wet walls. Suddenly, however, he cried out in fear. He had been lowered into a fast-moving stream of ice water. He was caught by the current and almost plunged under the water.

Louis gasped for breath and jerked hard on the rope. "Heave me up!" he yelled.

Francois was watching for signals, but by the time Louis's hard tug reached his student, it had lost force. Francois felt only a mild pull.

I guess he wants to stop," he said uncertainly.

The men held the rope steady.

"Where is he?" the professor asked.

Francois shook his head. "I can't see him at all."

Louis struggled to get his head above water. Finally, he grabbed the signal line and used it to pull himself above the water.

The sudden movement jarred Francois, who was holding the other end of the signal line. He yelled as he began to slide toward the well. Just in time, the professor grabbed his legs. Now they both heard it, a faint but unmistakable call from deep in the well.

"Help!" Louis cried.

"Pull!" yelled Gressly.

Everyone reached for the rope and began to pull. Slowly Louis rose above the water.

A new danger threatened him. As the men pulled at top speed icicles broke loose and jabbed at Louis like shards of glass. Louis kicked himself into the middle of the well to avoid the ice.

By the time he passed the icicles, shards of ice had frayed the rope on his chair. When it wore through Louis grabbed the signal line for his final hoist to the top of the well.

His companions cheered when he finally made it. Armand quickly offered Louis his coat. Louis's teeth were still chattering from his dunk into icy water.

Despite the dangerous trip he had just made, Louis was still enthusiastic about his discoveries. "Think what we have learned," he said. "Layers of ice in glaciers may tell us how much snow fell centuries ago."

The professor was unimpressed. "People are more interested in tomorrow's weather than they are with what happened ten thousand years ago."

"Finding out what happened in the past will help us predict the future," Louis said stubbornly.

It was time to quit work for the day. The men gathered their equipment and went to the hotel.

The next morning a blizzard raged on the mountain. Louis started to use the time to write in his journal, but his ink pen stopped writing. He examined it closely. "The ink froze," he said gloomily. "It can't get much colder."

He reached out to warm his fingers over a fire of coals. "I won't be satisfied with this year's work until we take measurements of the glacier in the Strahleck pass."

The professor's head jerked up. "Once this storm breaks we'll have a few days of good weather," he said warily. "But after that autumn will bring frequent storms. If we must go to the pass, we should go soon."

"We'll start the climb as soon as the storm passes," Louis replied.

The next morning when Louis pulled back the blanket at the door of the hotel he was greeted with silence. The storm was over. He rushed back to his companions to awaken them. "Let's eat," he said happily. "I want to spend the night in Grindelwald on the other side of the pass."

The men ate a hasty breakfast and set out. As the men climbed the mountain their spirits rose. The sun was warm on their backs and glistened off the snow as they carefully made their way up the icy slope of the glacier. All of them were roped together to prevent accidents, but when they came to snow almost three feet deep, they found themselves struggling.

The next problem after deep snow was a wall of packed snow. Jacob cut steps in the wall so they could climb up the final stretch to the top of the Strahleck pass.

The pass was covered with unmarked snow that was achingly lovely in the clear mountain air. Francois, touched by the beauty, asked, "Have human footprints ever marked this pass?"

"Never before," said Jacob. "We're the first."

Gressly suddenly laughed and grabbed a handful of snow, packing it into a ball between his huge hands. He let it fly at the professor, who ducked and reached for his own fistful of snow.

Now everyone dropped their packs and joined the fun. Soon balls of snow were flying in every direction. These men hadn't had such fun in years. They wrestled and rolled in the snow, yelling at the top of their lungs. Who cared? No one could hear them here.

Finally, everyone sprawled out on the ground to rest. After a few minutes of silence, Louis said, "We have time for one hour of measurements. Professor, you read the air pressure on your barometer and figure out the height of the pass. I'll find out how cold it is up here. I also want to gather some samples of the snow. Francois, look around. See if the chamois live this high on the mountain."

Everyone went to work. When they finished, they packed up their gear and roped themselves together for the climb over the pass. It would soon be dark and they still had a long way to go.

On the other side of the pass, they came to another glacier. But this glacier proved to be a real problem; it was split by a crevasse twenty feet wide.

"How deep is it?" Louis asked. His team crept cautiously to the edge.

"It appears to be bottomless," said Jacob. "Let's walk along the rim. Perhaps we can find some sort of natural bridge across it."

"We must find a way across," Louis said. "We don't have enough time to turn back."

After a long, cautious search, the men found a narrow bridge of snow that crossed the glacier.

"It's only two feet wide," Francois complained.

"Still, it's a way across," said Jacob. "I'll go first."

He stepped onto the snow. As he walked onto the bridge a chunk of packed snow broke loose and tumbled into the crevasse. Yet the walkway supported his weight, so he walked quickly over it to the other side.

Gressly followed, and after him came the professor. Each man that walked over the bridge, however, shook loose a little more snow. By the time Wahren, the porter, approached it, much of the bridge had fallen into the crevasse. Wahren hesitated at the edge.

"My pack is too heavy," he whined.

"I'll take it," Louis said impatiently. Already it was getting dark. In no way would he allow this man's fear to endanger everyone else's lives. "Go NOW!" he shouted to Wahren.

Slowly the porter stepped forward. He walked with short steps, an expression of sheer terror frozen on his face. As he neared the end of the bridge he lunged for safety, only to find himself slipping toward the crevasse.

Jacob and Armand yelled and pulled on the rope linking them to the porter, pulling him to safety.

Francois crossed the bridge next, springing across it as if there were no danger. Louis Agassiz was the last one to cross, but he had to cross the weakened bridge with two packs. Still, he proceeded, with slow, confident steps. At the other side, he dropped Wahren's pack into the embarrassed porter's arms.

Early that evening they reached the small village of Grindelwald. People in the village rushed forward to greet them.

"How did you get here?" they asked.

"Through Strahleck pass," Jacob said.

The villagers clapped their hands with delight. "Stay with us tonight. We will celebrate your victory."

Late into the night, the town was alive with activity. Everyone sang songs and told stories about what they had seen and where they had been.

As years passed, Louis became Europe's best-known scientist. One day many years after his trip to the glacier, Agassiz went walking with his children to the top of a grassy knoll. Alex was ten years old already, and his sisters, Ida and Pauline, were eight and four.

As he walked with his children, Louis talked to them about nature, bending down to pick grass, capture small insects, and examine rocks. Each child carried specimens of the day's walk as they reached the top of the knoll and paused to appreciate the view of the lake and mountains.

"The earth is full of the goodness of the Lord," quoted Louis from the Bible. Then he said to the children, "We must remember that nature is God's creation, his special gift to man. He has given us earth for a mansion, the blue sky for a roof, and green grass for a carpet."

Alex nudged a furry caterpillar with his toe. "There's a worm on our carpet," he said.

Louis laughed. He led the children back to Cecelia who was sitting in a chair under a tree. On the ground were a picnic basket and a blanket. Louis had even set up a small table for the meal.

Although it was bright and sunny, Cecelia was covered with a heavy woolen blanket. While Louis had become stronger and more robust through the years, Cecelia's health had declined. Every day she became weaker. Today Louis sat down on the grass by her side while the children ran off to play. Jacob Leuthold, Louis's guide on many expeditions, walked over to join them.

"Jacob and I are planning to make one final assault on the mountains," Louis said to his wife. "This time we want to climb the Jungfrau. If we can reach that summit I'll fulfill one of my biggest dreams."

"The Jungfrau?" Jacob was doubtful. "Even professional climbers consider that peak unreachable. I tried to climb it once but gave up. The last attempt ten years ago by climbers failed; six men died."

"This will not be mountain climbing for sport," Louis said firmly. "This will be a scientific expedition. I have never lost a man in the mountains because I refuse to risk lives. We will not attempt the climb until we have done everything we can to ensure a safe climb to the top."

Jacob picked up Louis's binoculars and looked at the Jungfrau. "The summit is as pointed as the spire of a church," he said. "We'll need five extra men as porters. The starting point should be the village of Grimsel."

"Let's go," said Louis.

## Chapter 9: Victory over the Jungfrau

In the dim light of early morning eleven men walked away from the village of Grimsel to trudge up the mountain. As the day began to brighten, so did their hopes for this climb. Then suddenly Jacob noticed blue windows in the ice ahead of them.

"Take care," he warned. "This ice is thin and opens into a well. Step carefully or you'll spend eternity in a cold, dark place."

He led them carefully around the holes, pointing out each window. Then he stopped. "This window has become clear," he said. "It must have frozen over a crevasse.

"What should we do?" asked Jean.

"We'll have to cross it," Jacob said. "We'll rope together and travel single file."

The thin ice groaned as they walked over it, but held firm.

They crossed the crevasse and climbed higher. Now fog settled around them, as the way became steeper. Jacob cut steps in the ice, but it was slow, dangerous work. Once a porter slipped as they hung on a cliff of ice. The others dug in their spiked boots and held on to his rope. Inch by inch they pulled him back until his feet were back on Jacob's steps.

That night they made camp on the broken ice at the top of the Jungfrau's glacier.

The next morning they had barely begun to climb before they met another challenge. This time a wide crevasse, almost twenty feet across, blocked their way. "Bring me the ladder," said Jacob. He swung the ladder across the crevasse to make a bridge, but it barely spanned the opening.

"Take off your packs and come across," commanded the guide. "We can swing those across later with the rope."

Jacob, Louis, Francois, the professor, and Armand crossed the ladder. But Wahren, the lead porter, and five other porters laid down their packs. They stood together, talking in whispers.

Louis realized trouble was brewing. "Cross over," he urged the men.

The six porters shook their heads. "We will go no further," Wahren said.

"We need you," replied Louis. "You must carry our food and equipment."

But despite his pleas, the porters hoisted their packs and turned back down the mountain. Jacob looked at Louis. "There goes our food," he said unhappily.

"But not our scientific instruments," Louis said. "We can still go up the mountain; we'll just have to limit our time on the peak to one day."

The five men walked forward until they came to a sheer ice wall. Again Jacob cut steps, pinning down the ladder to use as a handrail. They made it to the top and found themselves standing on a six-foot-wide snowy ledge. Jacob studied the mountain again to determine the best route.

While they waited, Gressly jabbed his ax handle into the ledge and was surprised to see it disappear into the snow. He dug around until he found it, but then stared into the hole the ax had made.

His eyes widened. "Move over to the wall," he whispered.

"Why?" Francois asked.

"We're standing on a ledge of packed snow," replied Gressly, "not rock. I looked into the hole the ice ax made and saw thousands of feet below us into the valley!"

Everyone crowded to the wall, and as they waited, pieces of the ledge broke away.

"Hurry," Armand told Jacob.

"This way," Jacob said at last.

The mountain rose in giant steps. They would climb one twenty-foot wall, rest on a flat ledge, then climb another wall. Every ledge they came to became narrower until finally, they rested on a shelf so narrow that the men could barely find room to stand.

Jacob told them to stop.

"What's wrong?" Louis asked.

"We're on the last ledge before the peak," Jacob said.

The professor sank to his knees. "That's it. I can't climb another step."

"There isn't room on the peak for more than one person at a time anyway," Jacob said.

Louis and Jacob removed their packs. Louis hooked his ice pick in a crack in the rock and pulled himself up to a toehold. With a push from Jacob, he made it to the top.

The Jungfrau did come to a peak, with a summit that was only six inches wide and less than two feet long. Louis slowly rose to his feet. He felt like he was standing on the top of the world.

He could see a hundred miles. Off in the distance he recognized the Matterhorn and Mont Blanc. Suddenly his eyes filled with tears. "Surely this is the most beautiful sight in the world!"

After Louis came down from the summit, the others took turns climbing up. Then, suddenly, it was time to climb down the mountain.

As they prepared for the descent Jacob said, "We must not allow our minds to wander from the task of climbing down safely. Often more lives are lost going down a mountain than coming up."

"It will be difficult to cast off the spell of the Jungfrau," Louis said. "That memory will be with me forever."

"Mountain climbing does change a man," Jacob agreed. "But watch your steps. You're still mortal."

Before they reached Grimsel, darkness fell. The moon rose, casting eerie shadows across the ice. They decided to go ahead anyway. It would be better to reach Grimsel in darkness than spend the night on the ice. As they hiked down, they heard the sound of a yodel. Soon a herdsman came to greet them.

"Stop and rest," the herdsman said. "Everyone in town watched your victory on the Jungfrau through the telescope. They sent me to welcome you back and bring you a bucket of warm milk."

The five weary climbers gathered around the herdsman for a drink of rich, fresh milk. Soon they felt refreshed enough to sing as they walked toward the village.

"Our task on the mountain is finished," Louis said to his men. "As soon as possible I'll present my findings to the Helvetic Association."

Louis had gathered much data about the ice age that he hoped to present to the Helvetic Association. This time, however, when he spoke to the famous scientists and teachers, they listened carefully.

"We traced the movement of eighteen boulders on a glacier," Louis began. "We made precise measurements and have proof that glaciers do move. We drove stakes into the ice and they moved too. The stakes in the center also moved faster than those at the edge."

The scientists nodded, impressed by the data.

Louis continued. "Studies in Poland, Italy, and England all provide proof for the theory of the great ice age. About 20,000 years ago the tropical vegetation of Europe was buried under a vast mantle of ice. Everything was stilled; plants, animals, lakes-even oceans. All over, creation fell silent. Rivers ceased to flow in the icy blast of perpetual winter; the sun's warmth was powerless over this deep invasion by the north. The only sound now was the thunder of crevasses as they opened across glaciers in a sea of ice."

Louis finished his speech. The members of the Helvetic Association stood and applauded. It was a most satisfying moment for the man who had left in disgrace years before.

At home, however, things weren't so wonderful. Cecelia was so weak that she spent most days in bed. Today she was propped up against pillows so that she would write on a small table positioned over her lap.

She held up a handful of mail as Louis came in to see her. "Letters!" she said happily, "from all Europe's great science centers. They want you to lecture at their universities. "

She read the return address on one of the letters. "The Royal Society in London has offered you a professorship. That makes three requests now; Lausanne, Geneva, and now London!"

She fished for another letter, then looked at it more closely. "This one is from Baron Humboldt!"

The baron's words were precise and complimentary. "I ask your pardon for criticisms in former letters to you about your theory of an ice age. I called that heresy at first, but now I know I should always keep an open mind toward your investigations."

Louis nodded. "Scientists today all seem to believe in a prehistoric ice age; I'm not the only one looking for signs of it. The British were the last to assent to the theory, but now even they agree with me. Everyone that is, except Charles Darwin. He still is as cold to my findings as the Aar glacier."

"Why is that so?" asked Cecelia. "Why should Charles Darwin reject the ice age?"

Louis chose his words carefully. "For many years Charles Darwin has been working on his own theory. He believes prehistoric man may have descended from apes. Darwin knows I believe in God's special creation of man and thus must reject his strange theory of evolution. So, because I chose to disregard his theory. Darwin will not support my views on the ice age."

Louis paced the floor restlessly. His work with glaciers was finished, yet he couldn't bear to sit idle. When he had been away from home Louis had depended on Cecelia to keep up with correspondence and everyday work. But as he became busier Louis had found it necessary to hire assistants; printers, student investigators, and artists to help him write the many books he published. At times twenty-five people worked for him. Now that Cecelia was ill, however, all that activity must cease or she would become even weaker.

"It's time for a change," Louis said. "I'll auction off the printing press, dismiss my assistants, and send away the research students. I'll even return the fossils I've borrowed."

"Why?" Cecelia asked. "Your work is your life."

"My work is finished here," Louis said. "It's time to find something else. Sir Charles Lyell from England just returned from America. He tells me people in the United States want to hear more about the ice age. He even arranged for me to lecture at the Lowell Institute in Boston. Cecelia, let's go to the United States!"

"Impossible," Cecelia said immediately. "I will not permit our children to become vagabonds because of science. They will stay here with me until you return."

"But what about making a new home for ourselves in America?" Louis pleaded.

Cecelia sighed at the indomitable persistence of this man. Arguing with him was useless. "Very well," she said, "after you have visited the United States and located a home for us, we will come and join you."

On the night before Louis's departure to America, he found himself too uneasy to sleep. He stood in the middle of his office, looking around. He had packed everything but the blue vase he had found in the fisherman's net years ago. As he examined the blue vase and started to wrap it, the door of the study opened softly. It was Alex.

Louis looked at his son fondly. "I was going to store this," he said, "but now I want you to have it."

"I found it a long time ago," he said. "They were once so common that fishermen broke them when they found them in their nets. Today we know that these vases were once made by lake dwellers thousands of years ago. Now this blue vase is a rare collector's item worth more than any fish."

Alex took the vase, tears filling his eyes. "I wish I could take you to America," Louis said, "but I think you should stay with your mother. You must help her while I am gone and remember your schoolwork."

Louis looked at his watch. "It's time for me to go."

Torches flickering outside drew Louis to the windows where he heard the sound of marching feet. Louis pulled back the curtain to see many people from Neuchatel who had come to say farewell. The scientist waved at them and told them he would be down soon. Then he went to the bedroom to kiss Cecelia.

"Monsieur de Coulon and the people are outside," he said as he brushed her cheeks with his lips. After he left she sighed, falling back wearily on her pillows.

Louis went downstairs and opened the front door.

"We have come to say farewell," Monsieur de Coulon said. "Neuchatel will never be the same without you. Our lives are less because of your departure."

"I will come back," Louis promised.

"No," de Coulon said shaking his head. "Switzerland is too small for a man with your vision. You will go to America where you will become even more famous. "

As Louis left the house, the people followed him. Everyone was sad that he was leaving. At the station, Louis gave his baggage to a man who hoisted it into the back of a horse-drawn coach. Then Louis said good-bye to everyone and stepped up into the stagecoach.

The coach jerked forward. "Farewell, my friend," Monsieur de Coulon said.

"God's speed to you," de Pourtales cried.

Louis was alone. He put his head down. Only with great effort did he manage to fight back the tears.

Chapter 10: A New Land

Louis stood by the rail on the deck of a three-masted sailing ship. He was on his way to Boston. As he breathed in the cool salt air, the captain joined him. "Is the naturalist enjoying the voyage?" he asked.

"Very much," Louis said. "All my life I have investigated mountain tops and glaciers. Yet, since I was a child I dreamed of crossing the ocean. I would very much like to have your permission to climb that," he said, pointing to the ship's rigging.

"Why?" the captain asked, startled at this man's boldness.

"I understand we are so far at sea that land cannot be seen in any direction," the scientist said calmly. Then he grinned. "I want to make sure of that from the highest spot on this vessel."

The captain smiled back. "We are so far from shore that we will not see land for another thirty days. But go ahead. Climb the mast. It's a spell-binding sight."

Louis climbed to the top of the ship, a dizzying height. From the rigging he looked as far as he could, but nowhere did he see land. He saw only sunlight sparkling over the waves of the Atlantic Ocean. Louis finally nodded and climbed down.

A month later the ship docked in Boston. Immediately Louis was forced to adjust to the fast-paced swarm of crowds of people going about their business. His baggage was unceremoniously dumped over the side of the ship where it landed at his feet. Louis looked around for someone to help with it, but everybody looked much too busy.

He spotted a man pushing an empty wheelbarrow.
"Will you carry my baggage?" Louis asked.
"Where to?" the man asked.

"The home of John Lowell in Pemberton Square," Louis said.

The man agreed to take the luggage and threw the bags into the wheelbarrow. Louis winced as the bags thudded against each other. Never before had his things been so mistreated.

Louis walked fast to keep up with the man-everybody seemed to be in such a hurry!

Finally, they arrived at the home of John Lowell. Louis paid the baggage handier and pulled the doorbell. A young man opened the door.

"You must be Doctor Agassiz," the young man said. "Do come in. I have been looking forward to our meeting. Already we have made arrangements for you to lecture about the ice age."

"You are John Lowell?" Louis asked.

"You expected an older man perhaps," Lowell said dryly. Then he smiled. "Science is a young profession in the United States. Until today science and the arts were considered luxuries that few could afford. Now, however, people have more time. They want to learn about nature and history."

"My first lecture will be about God's planned creation," Louis said.

"That sounds good." Lowell looked thoughtful. "If you can," he suggested, "write out an advance copy for the Tribune. They want to print all your lectures."

Louis looked surprised. "Will people come to lectures if they can read them in the newspaper?"

Lowell grinned. "You forget Louis Agassiz is world famous. Your lectures are so popular that already they are sold out. The first night you speak will already be to a crowd of 1500. We may even be forced to offer two lectures a night."

Louis looked unhappy. "I had hoped lectures would not take all my time."

"Of course not," Lowell assured him. "We want you to visit our country. You'll have time to travel; Charleston has already asked for you to speak there. And I think you will enjoy riding our rail system; many think our trains are the fastest in the world."

As weeks passed Louis did travel through the United States, filling his senses with the sights and sounds of this vast, young country. But although his restless nature felt much in common with the speed and power of America, something deep inside still yearned for the calm order he had left behind in Switzerland. His letters to Cecelia were colorful and frequent, but in them, she too sensed Louis's sadness.

"People here are friendly, " he wrote. "Workers wear clean shirts and go to schools just like rich men. Everyone takes pride in doing things for themselves; I have not seen any beggars on the streets or met a man without a job.

"Yet I am sad," he continued, "When I think of my children growing up without me. Please come to America; a sea voyage might make you stronger. We can make our home here in this vast, rich country. Please write me and tell me when you will arrive."

The next letter Louis received from Cecelia, however, did not refer to his request. Instead, she included the usual: pencil sketches of Alex and the girls, and chitchat about what they were doing. "People say I am an invalid," she wrote in spidery letters, "yet I still get out for walks. Alex's passion for collecting specimens continues -- you should see his room!"

The house Louis rented in America was alive with people. Any grief he might have felt about separation from his family was soon forgotten as students, scientists, assistants, and friends from Europe dropped by to visit. Even newspaper reporters frequently visited him. Rarely did he disappoint them by failing to provide new material for good stories.

"Now that you have finished your lectures," one reporter asked him this morning, "what will you do next, write another book?"

Louis shook his head. "I do not like to sit at a desk and write. Some of the books I started will never be finished because of that. No, I guess the next thing I am considering is the Harvard expedition."

An assistant entered the room, preventing him from saying more. "Two gentlemen are here to see you," the assistant said, "Professor Jacques Burkhardt and Francois de Pourtales. I said you were busy but they insist -- "

Louis jumped up, startling his assistant as he rushed to the front door. With a broad smile, he threw open the door to find his mountain climbing comrades. "My friends," he cried, embracing them. "What brings you here to America?"

Francois said lightly, "We couldn't stay away."

But the professor gave the real reason. "Europe is torn apart by war," he said. "Cecelia and the children are safe, but many people cannot find work in Switzerland and France."

"Why?" Louis said. He hated war. "Why must fighting tear apart so many lives? Even animals know better -- birds of the same species sing in the same key all over the world. Why can't human beings follow their example?"

The professor looked sad. "No one is more useless in war than a painter."

"Or a natural scientist," Francois added.

"You must stay here then," Louis said quickly. "Work for me again! I was just considering Harvard's expedition to Lake Superior as my next project. Now we can all go. I need your help."

Francois and the professor smiled. Trust Louis to find work when no one else could! "When someone is next to Louis even on a cold day he has less need of a coat," Francois said gently. "His smile keeps away the cold."

"Surely you both work hard enough to merit generosity," Louis said quickly. He clapped his hands. "I

must tell the cook to set two more places for supper."
When the cook heard about two more dinner guests he sighed. Two more tonight. Twenty-five last night. "There's nothing to serve," he said unhappily. "Last night I emptied the pantry."

"Many of my guests are war refugees from Europe," Louis said. "I simply can't turn them away. Check the pantry again. You'll think of something."

He did. When Louis and his guests sat down for dinner the delicious aroma of homemade soup tantalized their noses. Francois and the professor hadn't eaten a good meal for days, and this meal promised to be wonderful. They tried not to watch too closely as the cook dipped his ladle into the rich broth, and to reach for their spoons more quickly than they should. But ah -- it was marvelous. Seconds? Of course.

Only when the tureen had been emptied completely did Louis suggest something other than eating. "You must all come to the pool behind the house now," he said, "to see the giant leatherback sea turtle a ship captain brought me."

"I thought leather-backs were only found in the Caribbean," Francois said reluctantly facing his empty bowl and laying down his spoon.

"That's what most scientists thought," Louis said, "but this one was found near Cape Cod. You will be amazed at its size. A team of horses had to deliver it on a cart."

Louis stopped talking, silenced by the cook's strange behavior. His face was ashen and his eyes frantic as he motioned toward the soup, then pointed to the back of the house. Louis looked startled, then understood.

Francois missed the exchange. "Such a beast is certainly worth closer examination," he said, rising from his chair.

Louis cleared his throat. "Uh -- perhaps later."

Francois looked up, wondering what was going on, then recovered. Gracefully he commented on his host's fine meal. "Never have I eaten such fine soup," he said. "May I ask what was in it?"

Louis bit back a laugh. "Turtle, I fear," he said.

The conversation shifted to other things. "Harvard's expedition this spring will explore the wilderness area west of Lake Superior. The only people that live there are Ojibway Indians. They are supposed to be a peaceful tribe, but John Lowell says we can't be too sure of that."

Louis pushed back his chair and looked at his soup howl. "But then, maybe we can't be too sure of anything, can we?"

Chapter 11: Ghost Lake

    The expedition to Lake Superior included Francois, the professor, Louis, ten Harvard students, and three guides. They traveled up the river by canoe. Louis looked at a map. "We'll shoot the rapids through Sault Sainte Marie, then follow the northern shore of Lake Superior."

    He looked at the trees along the shore. "I wonder when we'll meet the Indians."

    After lunch, the explorers gathered samples of local plants. Then they were back on the river. Late that afternoon a student was the first to spot Indian teepees. He called out at once to Louis.

    "We'll go ashore," Louis said. "But we must proceed slowly. We don't want to startle them."

    The students beached the canoes. Then they all walked cautiously toward the teepees. A dog began barking and smoke drifted up from fires, but nowhere did they see Indians.

    "This village is deserted," Louis said.

    "The Indians must be fishing or hunting," a student said.

    Everyone was disappointed. Still, there were miles to cover, so all returned to the canoes. Later that day as the canoes rounded a bend in the river, the explorers finally saw what they had been looking for: Indians! They were standing upright in canoes, maintaining extraordinary balance in the fragile vessels as they threw fishing nets over the stream. Farther and farther went the nets in precise arcs from the Ojibway's hands.

    The explorers were fascinated. "Are they hostile, do you think?" a student whispered.

    "I can't tell," Louis said. He ordered the men with paddles to direct the canoes to shore.

By now the Indians had seen the party of white men, and with strange-sounding cries signaled each other to pull in the nets. In minutes they were racing downstream, away from the foreigners. Soon they had vanished. By the time the explorers reached shore, the only sounds that could be heard were those of rushing water and rustling trees. The men stood on the beach, breathing heavily. They were alert and watchful, but they did not reach for weapons.

"They're gone," Francois said.

"Maybe they're planning an ambush," a student suggested.

"We'll return to the water," said Louis, "but proceed with caution."

After the canoes were back in the water and the trip was resumed the explorers spotted another Indian. He was alone, squatting down on shore. When he saw the canoes he rose, standing with his arms crossed.

Louis held up his hand. "Beach the canoes," he said.

They pulled ashore upstream from the Indian.

"Where are the other Indians?" whispered a student. Louis had talked to many kinds of people before, but never an Indian.

"Maybe he wants a powwow," said Louis.

"Or a scalp," a student said.

Louis walked toward the Indian. The Indian watched him without expression. Then Louis raised his hand, palm out. "How," he said.

The Indian stood still as a statue. His eyes didn't move.

Louis felt foolish holding up his hand. He lowered it and then said, "Hello. We mean no harm. We are explorers. Do you understand English?"

"Certainly, monsieur," the Indian said with a heavy French accent. "French, too."

Louis was surprised.

"You did not expect to meet a savage who spoke English, perhaps?" the Indian said. "Why not? French trappers and English gold-seekers have passed this way many times. They have many times used us as guides and hunters."

"We're not trappers or gold-seekers," Louis said. "We are scientists who wish to learn more about this land and its inhabitants. Do you understand?"

"You are what my people call a wise man," the Indian said.

"The work we do would go faster with experienced guides and hunters," Louis said.

"That can be arranged," said the Indian.

"How are you paid?" Louis asked.

"Our people have kept their ways even after the white man came. But silver and gold can be used even by the Ojibways."

The Indian gave a signal, a small but precise movement of his head. Suddenly Indians popped out everywhere from underbrush by the stream. They glided silently up to shore.

Louis waved to his own companions to come close.

"The Indians will be our guides," Louis said.

That night the scientists and Indians made camp together. The explorers watched as Indians spread fish out on wooden sticks and smoked them over the coals of a fire.

Francois filled his mouth with berries. "This trip has turned out to be a real vacation. No longer must I look forward to falling rock, or cracking ice, or even going to sleep to wake up freezing to death."

Louis nodded his head. "This is a land where nature is rich."

Professor Burkhardt was sketching a fish that was lying on a white strip of birch. The Indians had tried to throw it away because it was inedible, but Louis had stopped them.

"The Indians can't understand why I want fish that can't be eaten," Louis said.

"You're In danger of loosing your title of wise man," the professor said, smiling. He showed Louis his drawing. The fish was sketched in full color.

Louis nodded, satisfied. He took the fish and plopped it into a small wooden keg filled with alcohol.

"Alcohol preserves the flesh, but not the color," Louis said. "Without your drawing, we'd never be able to remember what it was when we got back to Boston."

Louis called to the others. It was time to summarize the day's findings. "You are learning more here in a day than you would in months from books," Louis said. "Most of you are from the city where you have become blind and deaf to things in nature. My wish for you is that you might regain full use of your senses. I want you to see, hear, smell, and above all to know the world around you."

Louis pulled down his makeshift blackboard, a black canvas cloth that was on a roller like a window curtain. "Here's what our explorations teach us," he said, positioning a piece of chalk on the cloth.

"Today there are five Great Lakes, once there were six. In prehistoric times this sixth lake covered a vast area west of Lake Superior."

"How do you know this?" asked a student.

"The evidence is all around you," Louis said. "Fossils and geological signs help us to map out the precise location of the ghost lake-even to date its existence. This lake was born in the Pleistocene era when it was formed by melting ice from the last great ice age. This is what it looked like."

Louis sketched a crude map of the United States and Canada, then included the outline of the missing lake. It was an enormous expanse covering the states of Minnesota, North Dakota, and Canada's Manitoba.

"The lake was 700 miles long and almost 300 miles wide," Louis said. "It lasted almost a thousand years but disappeared in time because most of it gradually drained into the Hudson Bay when the glaciers melted. It did, however, leave behind many small lakes."

"Glaciers once covered most of the earth," he continued. "When they melted they changed the whole face of the earth. As ice melted 20,000 years ago, it left behind most of the world's lakes."

The expedition pushed on in the morning. Finally, they arrived at Fort Williams, an old fort in the Hudson's Bay Company chain. The fort was abandoned, but outside it were log cabins for people who still lived in the area. As they walked up to the trading post a man walked out to greet them.

"Dr. Louis Agassiz?" the man asked.

"Yes," said Louis. "These are my men and our Indian guides."

"My name is Mackenzie," the man said. "I have mail for you."

He gave Louis some letters. Louis took those addressed to himself and handed the rest to Professor Burkhardt.

"We thought Fort Williams was abandoned," Louis said to Mackenzie.

"It's seldom used," Mackenzie said, "but the town around it is growing. Copper mines are opening south of here and on Isle Royale."

Louis looked through his mail. His eyes showed disappointment. "I had hoped for a letter from Cecelia telling me she would join me in America," he said in a low voice. "Instead there's a letter from her brother. I wonder what it means?" He walked away from the others, suddenly very anxious to be alone.

Louis sat on a wooden bench under a tree and fingered the letter from Alex Braun. Then slowly he opened it and began to read.

"Since your last letter," Alex wrote, "Cecelia grew weaker and weaker. She who has had so many trials in life has finally found rest. Cecelia is dead." Louis's hands began to shake. Tears filled his eyes as he continued to read. "Your mother is caring for the girls, but Alex is with me. He says he wants to come to you as soon as possible."

Louis sat like stone, grief breaking inside like waves on a shore. The professor waited a few minutes, then walked to Louis, putting his hand on his shoulder.

"My friend," he said quietly.

"It's Cecelia," Louis said. His voice failed.

"She has died?" the professor asked gently.

Louis nodded.

"Who has the children?"

Louis cleared his throat and swallowed hard. "Ida and Paulene are with my mother. Alex is with the Brauns. I must send for them right away."

The professor waited for a few minutes, then spoke firmly. "Your home in Cambridge is hardly a proper place to raise two young girls, my friend."

"You are right," Louis said after a pause, "but I will send for Alex! He's fourteen and old enough to learn the ways of another country."

Chapter 12: Nature's Librarian

Alex was in America! He stood stiff and erect at the dock in Boston. Louis gave a joyous cry when he saw his son, then ran to embrace him. He gazed at the boy in obvious pride. Alex was a serious, well-mannered boy who looked and acted like an adult.

"Professor Felton has already invited friends to his home to meet you," Louis said. "It's supposed to be a party, but I'm afraid it might seem dull to you. Later we'll do things together though, like riding a train and visiting the beach. Would you like that?"

"Certainly, Father," Alex said politely.

At Professor Felton's home Louis introduced Alex to his Harvard friends. Alex responded to the others with a stiff German bow and clicking heels. They found his formality strange and had to hold back the urge to laugh. Alex sensed the other guests' amusement but found it baffling. Gradually he became restless and uncomfortable. Then Professor Felton approached with a lady. "Alex -- and Louis -- meet my sister-in-law, Miss Elizabeth Cary."

Elizabeth gave a gift-wrapped package to the boy.

Alex lit up, shedding his formal manners. "Oh, thank you!" he said happily. Then he tore open the paper.

On the other side of the room, Professor Felton and Louis continued to talk. "To display all the new specimens that you brought back from Lake Superior, I expect Harvard will have to build a museum. Do you agree?"

Louis was watching Elizabeth.

"Don't you agree?" Felton persisted.

"What?" Louis asked.

"Harvard must build a museum for your collection," Professor Felton said, "do you agree?"

"I suppose so." Louis was still watching Elizabeth who was helping Alex with the present she had given him. It was a wooden puzzle that could only be put together in a certain way.

Felton followed Louis's gaze. "Your son is a handsome lad," he said.

"Miss Elizabeth Cary," Louis said thoughtfully.

"What about her?" the professor asked.

"My son seems to be taken with her," Louis said. "She is a beautiful lady." He walked over to his son.

As weeks passed, Elizabeth escorted Alex all around Boston. She and Louis became good friends, too.

One day, all three of them went to the beach. The weather was cool. They were the only ones walking on the shore. Alex ran to the water's edge. He was dressed in casual clothes; quite a change from the formal dress he wore when he first arrived in America. His face was streaked with mud as he fished out a mat of seaweed.

"Alex has learned American ways quickly," Elizabeth said. "He has also learned how to play again."

"He would not have learned so quickly without you," Louis told her. "Alex is very fond of you."

Elizabeth looked down shyly. "It must have been awful for him; a stranger in a strange country. Still, he's happy now, and so am I."

"So am I," Louis said softly. "Alex loves you, but he is not the only one. I too love you very much. And Alex needs a mother!"

Elizabeth was speechless.

"Not only do my children need you," Louis said, almost tripping over his words, "I need you too. Elizabeth, will you marry me?"

Elizabeth stared at him.

"Say yes," Louis said impatiently. "I must have you and my children. Then life will be complete."

Elizabeth eventually found her voice enough to accept Louis's proposal, and in time, Louis sent for his daughters so they could be with him and Elizabeth in America.

They were married in a chapel. On one side sat the Carys, Elizabeth's large and famous family. On Louis's side of the chapel sat only Alex, Francois de Pourtales, Professor Burkhardt, and John Lowell.

The night they were married Elizabeth moved into the Louis Agassiz home. As she unpacked her things, she opened a cupboard to put in her boots. What she saw inside made her scream in terror. A snake was squirming over the floor.

Louis ran into the room. "What's the matter?" he asked.

Elizabeth just gasped and pointed at the snake.

Louis looked down, then showed concern. "Did you find only one? I carried five up here last night in a handkerchief. Where are the others, do you suppose?"

"You'd better find them, " Elizabeth warned, walking to the door.

"The prettiest one seems to have escaped," Louis said to her back. "And I did so want you to see him."

His answer was another scream. Elizabeth stood in the hall with her back against the wall.

"An alligator, in the bathtub," she said weakly.

Louis looked ashamed. "He was eating turtles in the pond and I had to remove him until I could find him another home." He looked up to meet his bride's suspicious stare.

"What else is lurking around?" she asked.

"Well," Louis said slowly. "There's a family of opossums in the garden, turtles in the pond, a tame hawk in Alex's room, and rabbits under the backstairs. I think that's it."

But he wasn't sure. Perhaps he had forgotten something.

"All animals out of the house," Elizabeth shouted. "I will not live in a zoo!"

Louis looked at her uncertainly.

"This very night! Otherwise, I will not sleep under this roof."

Louis obeyed, reluctantly. A farmer near the city agreed to care for the animals, and he was sent for immediately.

But Louis had forgotten one animal. In the cellar lived a large furry bear named Bruno, who currently was sniffing around a wooden barrel. Louis was supposed to feed him, but because of the wedding, he had forgotten. Now the animal's stomach grumbled and growled, but there was no food. Finally, Bruno gave up and laid down to sleep.

## Chapter 13: Desolate Theory

The next day Louis and Elizabeth entertained Louis's friends. He explained who they were. "Every Sunday afternoon this unofficial club has been meeting to discuss the important issues of the day." Then he introduced them to his wife.

"You already know Henry Wadsworth Longfellow and James Lowell," he said as the two men entered. "With them is Henry's son Ernest. He is one of my students."

Two other men arrived, Dr. Oliver Wendell Holmes and John Greenleaf Whittier. Whittier asked if Thoreau was coming.

"He's already here," Louis said, "although I suspect he'd rather be at his cabin at Walden Pond."

About this time Bruno woke up and began to prowl around. He had gone two days without food and he was hungry!

Elizabeth spoke to her guests. "I'm convinced I've married the most popular man in the world. France offered Louis the chair of paleontology at the Museum of Natural History in Paris."

"That's Professor Cuvier's old post," Louis said. Elizabeth continued. "Louis Napoleon even offered him twenty thousand dollars a year and a seat in the senate if he would accept."

"Harvard can hardly meet that offer," Lowell said. "Yet we are anxious for you to become a permanent member of our staff. It would be difficult to lose you."

Louis held up his hand. "No more talk; I'm staying," he said. "I would rather have the gifts of free people than the patronage of an emperor."

Bruno had managed to work his way out of the cellar and to the top of the stairs. Now he pawed at the door at the top of the steps until he managed to break it open.

Just as the cook was carrying a large roast out of the kitchen and into the dining room, the bear spied the meat.

"My poetry doesn't rank with Mr. Whittier's," said Holmes, "but here is a poem in honor of Louis's decision to stay in America."

He cleared his throat.

God bless the great professor

And the land his proud possessor --

The cook screamed. Bruno charged to the table and clamped his jaws over the roast. The terrified guests jumped up and ran out the door.

Louis slapped the bear on the nose. "Naughty Bruno," he said.

He pulled the roast from the bear's jaws and used it as bait to lead Bruno to the back porch. Then he examined the mauled roast, decided it couldn't be used for dinner, and tossed it into the backyard.

Louis returned to the dinner table. All the guests were dabbing at their clothing with napkins.

Louis apologized.

"Is the animal dangerous?" Longfellow asked.

"No," Louis assured him. "I forgot to feed Bruno yesterday before I left. The poor animal was probably starving."

"You mean that bear was in the cellar all night?" Elizabeth said, warning Louis with her eyes.

"We'll send for the farmer straight away," Louis said meekly.

Doctor Holmes decided to change the subject.

"Louis," he said, "have you found evidence in America for the ice age?"

Louis nodded. "Everywhere I look I see evidence. In my backyard is a fence made of boulders rolled smooth by glaciers. The soil around this house is made up of pebbles mixed with rock; remnants of glaciers.

"There's other evidence as well. Fossils in America are plentiful. The ice destroyed many plants and animals which left their mark in fossils."

Ernest Longfellow asked if Louis had read Charles Darwin's book yet.

Louis had read the Origin of Species. "Yes," he said.

Ernest became excited. "Darwin says all these fossils prove his theory of evolution," he said. "Plants and animals died, only to be replaced by stronger, better ones.

"Darwin also believes human beings are a product of evolution," Louis said gently.

"What do you think?" the student asked.

Louis sighed. "What I think matters little," he said. "Let the facts decide what is true, not public opinion. No one may use nature to prove his own views. She will force us back to truth when we wander." Louis looked at Ernest. "I do regret that so many young people find Darwin's ideas so exciting. How much better it would be if they would stick to careful, scientific investigation."

"Then we should investigate the matter!" shouted Ernest.

"Ernest is wild to go with you on your trip to South America," the boy's father said, "but he is too young for the trip."

"Ernest does have a point," Louis said. "The theory of evolution should be tested by the fire of truth. On this trip to South America, I had intended only to follow up on some work done by Baron Humboldt. Now I know I should do some first-hand research into the places Darwin explored for his theory of evolution. I must go not only to South America but also around Cape Horn to the Pacific."

"You must go to the Galapagos Islands," Ernest said. Louis nodded. "When I come back, I'll give my views on the theory of evolution."

Louis outfitted an entire ship for his journey. The emperor of Brazil welcomed Louis to South America and arranged for Louis to travel up the Amazon. The ship then sailed to the tip of South America to the Strait of Magellan where Louis explored a glacier at Glacier Bay.

After that Louis directed the ship around the tip of South America into the Pacific ocean toward the Galapagos Islands.

Years before Darwin had studied animals on the islands, believing them to be similar to prehistoric animals that once roamed Europe. Great turtles lived there, as did giant red lizards called iguanas.

Louis used the ship as a school, telling the students who had accompanied him about the islands.

"The Galapagos are about six hundred miles off the coast of Ecuador. They are very close to the equator," Louis said. "Many unusual creatures live there: giant tortoises, flamingos, and five-foot-long iguanas."

"Land ahead," a sailor called. "It's the islands."

The captain told them they would drop anchor first by Albermarle Island. "You can see it now," he said. "It's the one with the volcanic cone."

Louis called to his student assistants. "Lower the boat. We'll go ashore to explore."

They made camp on the island in a cave where lava from the volcano had once flowed. The studies Agassiz made proved the islands were new in comparison to the age of the earth. They had only recently been produced by volcanic action.

The islands were filled with strange and wonderful creatures, however. Louis and his students were busy collecting specimens from dawn until dusk. By the time the ship was ready to leave the islands, almost three hundred barrels of specimens were crowded into its hold.

A gun went off, signaling the explorers to return to the ship. Sailors had captured a giant tortoise. One sailor sat on the turtle's back, baiting the animal into moving with an apple suspended by a string from a pole. Slowly the tortoise lumbered over the wooden deck.

The sailors had also caught a shark. Louis inspected it and said to his students, "Sharks have highly specialized teeth and muscles. According to evolution, this fish is ancient; yet look at how well-developed its structure is."

"You do not accept Darwin's theory?" a student asked in amazement.

"The facts say something else," Louis said. "In retracing his voyage, what I see in nature is not the same as what Darwin saw."

When he returned to the United States, Louis went on record in opposition to the theory of evolution. He was one of the first scientists to do this, based on religion as well as science.

Louis talked about his voyage. "I have just traveled to all the places Darwin went to study his theory. I can say only that evolution is a desolate theory, using the laws of matter to explain away all the wonders of creation. It is a system that rejects God, substituting for our Creator only the impersonal, chance action of physical forces.

"Darwin assumes a transition between man and lower creatures. I cannot.

Everything in nature proclaims the God we know, worship, and love. I agree with the British statesman Disraeli who said, 'I am on the side of the angels.'"

Although Louis was born in Switzerland, he made the United States his home. In 1860 he became a U.S. citizen.

During his lifetime he received many honors, the greatest of which was the choice his son Alex made to follow in his father's footsteps. Like his father, a spirit of adventure and a firm commitment to truth possessed Alex.

Many students of science in the United States worked hard so they would be accepted into Louis Agassiz's classes, for he was considered to be one of America's finest teachers.

One reason why students enjoyed his classes so much was that Louis avoided teaching in the classroom in favor of field trips to museums or trips to places where nature could be studied firsthand. Often Louis invited students to accompany him on voyages or expeditions, and even had an island near Boston set aside for students to use as an outdoor classroom.

    Even after he became a citizen of the United States, Louis continued to travel all over the world. He went to South America more than once and visited Switzerland. Yet when he died on December 14, 1873, he was buried in Boston. Home to this world-famous man was his adopted country.

Chapter 14: Louis Agassiz Today

Louis Agassiz is remembered today as a great man of science who made many important discoveries. Although he had much success, Louis never claimed credit for it. "God wrote the books of nature," Louis said, "I am only His librarian."

Louis Agassiz was devoted to the truth of the Bible and was convinced that the great flood described in Genesis was responsible for the death of plants and animals that later were recorded in fossils.

Louis's fossil studies won him a reputation for being one of Europe's most accurate observers. As a student, he had already identified more than three hundred kinds of fossil fish, yet his books did more than describe them. He made the ancient seas in which they swam come alive.

His work on glaciers, although controversial at first, eventually forced the entire scientific community to accept irrefutable evidence that not only did glaciers move, but they had once covered the earth during a long wintry ice age.

When Louis Agassiz died in 1873, his friends fashioned a tombstone for his grave from a boulder found on the Aar Glacier. Yet his name is still heard today. The ghost lake, or sixth Great Lake he discovered near Lake Superior was named Lake Agassiz in his honor. And in 1915 Louis Agassiz was elected to the Hall of Fame for Great Americans.

The Bible taught man was only a little lower than angels. Some scientists believed man was only a little higher than apes. Louis Agassiz, like Disraeli, steadfastly maintained his own position "on the side of angels."

Other Science Books by John Hudson Tiner
Biographies of:
- Samuel F. B. Morse
- David Ligingstone
- Isaac Newton
- Robert Boyle
- Louis Pasteur
- Johannes Kepler
- James Clerk Maxwell
- The Ghost Lake -- Biography of Louis Agassiz

High Interest Books about Sciece
Perfect for Home Schoolers
Great as Supplementary Reading:
- Into the Unknown -Treasures of Nature
- Exploring the World of Physics
- Exploring the World of Biology
- Exploring Planet Earth
- Exploring the World of Mathematics
- Exploring the World of Chemistry
- Exploring the History of Medicine
- Exploring the World of Astronomy
- Exploring Physics
- Exploring Biology
- Exploring Earth
- Exploring Medicine
- Exploring Astronomy

John Hudson Tiner is an educator and author of more than 100 books. He brings a wide range of facts together to tell the exciting and interesting story of great breakthroughs in science. Although written to be entertaining, his non-fiction biographies have a strong educational content, too.

He says, "I especially like to write about the important events that have changed the world. After I finish the research, a magical moment occurs when the story takes over. The characters come alive. No longer am I a writer. I become a time traveler who stands unseen in the shadows and reports the events as they take place."

Check out these Kindle eBooks:

Crime, Action, and Adventure
    The Big Haul
    Shouldn't I be Missed
    Shouldn't I be Dead
    Deceptive Diamonds

For teen boys:
    Seven Day Mystery

Something Different
    The Secret Garden on a Diet [Annotated]

Detective Jeannie Bishop Mysteries by J. D. Tiner
    Matty is Missing
    Baby Rescued Mother Missing
    Missing Motive
    Missing Presumed Dead

www.ingramcontent.com/pod-product-compliance
Lightning Source LLC
Chambersburg PA
CBHW070350230526
45471CB00006B/2500